STALLS, SPINS, AND SAFETY

McGRAW-HILL SERIES IN AVIATION

David B. Thurston, Consulting Editor

AVIATION CONSUMER • *The Aviation Consumer Used Aircraft Guide*
MASON • *Stalls, Spins, and Safety*
RAMSEY • *Budget Flying*
SACCHI • *Ocean Flying*
SMITH • *Aircraft Piston Engines*
THOMAS • *Personal Aircraft Maintenance*
THURSTON • *Design for Flying*
THURSTON • *Design for Safety*
THURSTON • *Homebuilt Aircraft*
WHEMPNER • *Corporate Aviation*

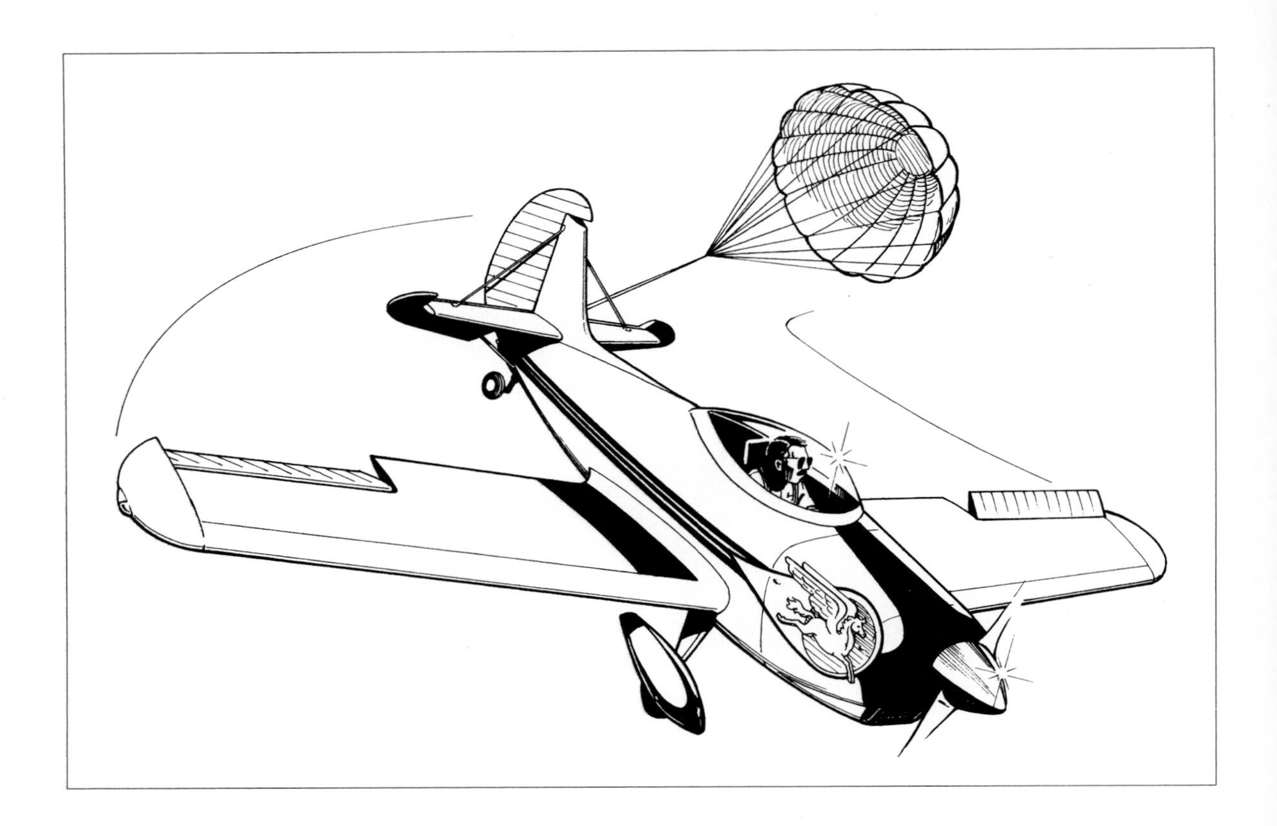

Sammy Mason

STALLS, SPINS, AND SAFETY

McGraw-Hill Book Company

New York St. Louis San Francisco Auckland Bogotá Hamburg Johannesburg London Madrid Mexico Montreal New Delhi Panama Paris São Paulo Singapore Sydney Tokyo Toronto

CAUTION: The practice of deep stalls and spins as described in this book should be attempted only under the supervision of a qualified aerobatics instructor. The author and publisher emphatically warn against unsupervised practice.

Library of Congress Cataloging in Publication Data

Mason, Sammy.
Stalls, spins, and safety.

(McGraw-Hill series in aviation)
Bibliography: p.
Includes index.
1. Airplanes—Piloting. 2. Spin (Aerodynamics)
I. Title. II. Series.
TL710.M38 629.132′52 81-18626
ISBN 0-07-040696-0 AACR2

1 2 3 4 5 6 7 8 9 0 VHVH 8 9 8 7 6 5 4 3 2

ISBN 0-07-040696-0

The editors for this book were Jeremy Robinson and Celia Knight, the designer was Elliot Epstein, and the production supervisor was Paul Malchow. It was set in Century Schoolbook by University Graphics, Inc.

Printed and bound by Von Hoffmann Press, Inc.

Illustrations by Wendell Dowling.

To Wanda Lee, the mother of six flying boys and two wonderful daughters.

And to my son Tony, who departed on eternal wings.

CONTENTS

FOREWORD

I was delighted when I was invited to write the foreword to Sammy Mason's very down-to-earth thesis on aviation's oldest and most fearsome maneuver, the spin.

Spins have been cussed and discussed since the first airplane crashed because of one. Although this phenomenon was first described as a "diving spiral," the more popular term "death spiral" was used until some aeronautical genius finally discovered what the devil was doing. This led to the development of spin-recovery training.

In 1926, Congress enacted laws governing commercial aviation, and the Aeronautics Branch of the Department of Commerce was created. Its primary functions were to establish procedures for licensing aircraft and pilots and to develop air traffic rules and regulations for all to abide by.

As part of the flight examination in those early years, pilots were required to demonstrate their skill by spinning the aircraft two turns in each direction, while being observed from the ground by the area inspector. Spin training and demonstration prior to licensing were discontinued many years ago, but they continue to be frequent topics of discussion, especially after someone we know has spun in.

The importance of spin training is a source of disagreement among flying schools and flight instructors. Flight instructors of the old school, who are still in business, encourage their students to accept stalls and spins as part of their primary flight training. Those who receive spin training are safer and more confident pilots, able to advance more rapidly through their instruction.

Because full stalls and spins are no longer required by the FAA, in most cases they are neither taught nor even discussed. It seems as if flight training has become less demanding, despite the fact that aviation requires such a great concentration of mental and physical skills. This softening of flight training is evident, as most of today's pilots lack the advantage of spin training and the discipline of recovering from other unusual attitudes. One wonders what can be expected of a pilot who has not had training in recovering from unusual maneuvers if the airplane is suddenly caught in severe turbulence and falls into a spin at low altitude. It can and does happen.

Sammy Mason is a member of the previously mentioned old school of flight instructors and is one of the most experienced, with the ability to impart his knowledge to others. I first met him in 1933, at the old Telegraph and Atlantic Airport in East Los Angeles, California. We learned to fly by the saturation method—several instructors and exposure to many different flight maneuvers. Sammy's career has been long and varied, and I'd like to highlight a few areas.

- In 1941, the late Tex Rankin noticed Sammy's ability in aerobatics and as an instructor and asked him to join the Rankin School of Aerobatics. This soon became known as the Rankin Flying Academy when the U.S. Armed Forces prevailed upon Rankin to start one of the first civilian contract primary flight schools. Sammy became a squadron commander in charge of "Squadron X," a group of instructors who worked only with cadets with learning problems.
- Following World War II, Sammy toured the United States giving aerobatic demonstrations. He used his converted Stearman trainer to perform in the first postwar National Air Races, held in Cleveland, Ohio, and went on to receive worldwide acclaim for his innovative routine.
- In 1950, Sammy was hired as a jet aircraft test pilot by the Lockheed Aircraft Company. He was invited to join the exclusive fraternity of test pilots under the world-famous aircraft designer C. L. "Kelly" Johnson of the Lockheed "Skunkworks." Sammy became Lockheed's authority on stall/spin testing, conducting tests on such aircraft as the T-33 and T2V-1 jet trainers and the F-94C all-weather jet fighter. Sammy retired from Lockheed in 1973, returning his skills and instructive ability to commercial aviation.
- In 1976, Sammy received the highest recognition for his instructional achievements by the National Association of Flight Instructors.

These examples provide only a glimpse of this wise and talented aviator whose vast experiences are presented in this book for all to read. The contents are of a serious nature, but you will also enjoy the humor for which Sammy is noted. You will sense the excitement of his experiences and be instilled with an everlasting respect for flying.

I consider *Stalls, Spins, and Safety* the finest book on the subject that I have read. Having known and worked with this man for nearly one-half century, I cannot say enough in my endorsement of what he has written. Read it carefully, and have a safe and sane flying future.

Tony LeVier

PREFACE

I was cautious as I started the spin investigation of a light military airplane. I made several spin entries, permitting the airplane to rotate one-half turn, suspiciously looking for indications of trouble.

Soon I was allowing the airplane to carry me through several slow and easy rotations. The nose was well down, stick and rudder forces were positive and firm. The recoveries were nearly instantaneous.

Although an almost identical airplane had killed a fellow test pilot and very close friend in a flat spin that brought him to the ground, I was beginning to feel secure as I spun in apparent safety.

I experimented with the controls as I tried to excite a flatter spin mode—without success. Then I applied just a trace of power. Suddenly the bird ruffled its feathers and revealed its disposition. The speed of rotation increased so dramatically that I had difficulty with visual cues. The nose lifted to the horizon.

I immediately attempted to reject the experience by shoving hard rudder against the rotation. It felt disconnected. The stick forces were nil as I shoved it to the forward stop. I tried several combinations of controls without success. Lifting my hand from the throttle, I reached forward and pulled a T handle to deploy the spin chute. It seemed like an ignored gesture. The altimeter swept through 3000. The spin continued several more turns. It was beginning to look as if I was going to have to bail out, my only chance for survival. Yet my buddy was killed, struck by the tail when he tried.

Suddenly the airplane pitched straight down and the spin stopped. The recovery was so abrupt that I was thrown hard against the canopy. I pulled another handle and jettisoned the chute while offering a prayer of thanksgiving.

The violent recovery had buckled the fuselage just aft the cockpit, and I had to stand on the rudder to maintain yaw control. I flew the airplane back to the base and enjoyed the happy squeal of the tires as the runway acknowledged my safe return.

SPINS—A LONGTIME PROBLEM It is the test pilot's job to goad an airplane until it reveals its true spin character. Airplanes that are approved for spins have been spun under the worst possible circumstances and found to be safe.

Under the supervision of a qualified instructor, spin training can be conducted safely. It is important to seek out a knowledgeable instructor because the Federal Aviation Administration (FAA) does not require that an instructor be tested concerning knowledge of spins.

Selecting an airplane that is approved for intentional spins, having it configured properly, and following the recommended spin recovery procedures are also important.

The objective, of course, is to train pilots to recognize spin-inducing situations and train their reactions so that they will recognize incipient spins and recover from them instantly. There is little value in continuing a spin beyond the stage where it is fully recognized as a spin. Recognition and recovery can usually be accomplished within two turns of rotation.

Most accidental spins occur at low altitudes (see the accompanying table, from SAE Paper 760480). However, safety demands that instruction in situations leading up to inadvertent spins be accomplished at a safe altitude. It takes very little altitude to recover from an incipient spin, but the training that develops the skills and reactions for quick and positive recoveries must be accomplished safely lest the cure become more hazardous than the ailment.

Frankly, I would be greatly concerned if the FAA suddenly ordered a return to spin training. Because of a general lack of skill in spin training

STALL/SPIN ACCIDENTS BY PHASE OF FLIGHT, 1965–1973
(Group of 31 single-engine aircraft, crop control excluded)

Phase of flight corresponding to first accident type	*Stall or spin as a first accident type, percent (Fatal accident percentage data in parentheses)*	
Takeoff	26	(15)
In flight		
Normal climb/cruise/descent	8	(11)
Acrobatics/low-pass/buzzing	18	(23)
Unknown/other	22	(32)
Total	48	(66)
Landing		
Traffic pattern	5	(7)
Final approach	11	(7)
Go-around	9	(5)
Total	25	(18)
Unknown/other	1	(2)
Total for all phases	100	(100)

Source: "Statistical Analysis of General Aviation Stall/Spin Accidents," SAE Paper No. 760480, table 4.

Spins—a longtime problem.

among flight instructors, I believe the number of spin accidents would increase.

MENTAL METAMORPHOSIS Safe flight starts with a proper mental attitude. If our attitude is proper, we will seek out the best possible flight training and try to become as proficient and as safe as possible.

However, even with good training and sufficient knowledge, each flight presents its own set of problems and challenges that demand a sober and thoughtful approach. Too often a preflight will include everything but a check on ourselves. Are we really fit to fly?

It is illegal to fly during times of physical deficiencies, but there are no rules that govern us during times of mental or emotional stress. Many accidents might be traced to the death of a loved one, a threat of divorce, or a fight with a spouse at the breakfast table.

What can be termed as "temporary insanity" is the catalyst of many accidents. Emotionally charged situations short-circuit a person's ability to think and reason.

During World War II, a flight instructor, upset by his student's sloppy flying and his delay in turning crosswind after taking off, grabbed the controls away from him and jerked the airplane into a steep turn so rapidly that the Ryan PT-22 snapped into a spin. The resultant crash killed the instructor and severely injured the student.

In his right mind, the instructor would never attempt such a thing. He knew the probable result of such rough flying and the extreme hazard of a low-altitude spin. Yet, in a moment of uncontrollable rage, he became suicidal.

Confidence is a necessary attitude for a pilot. However, false confidence can create an illusion of assurance and well-being that feels like the real thing. I have witnessed this in people who have been successful in other pursuits. Sometimes a confidence that relates to other skills tends to carry over into the unrelated field of flying.

Often, people with exceptional skills and intelligence in other fields act as though they were mentally deficient in the judgmental aspects of flying. Unfortunately, they not only have a high accident record while at the controls of an airplane, but they have been influential in tainting the judgment of others who are flying as pilots in command.

A common complaint of executive pilots is that their bosses tend to push them into situations that their better judgment tells them to avoid. Their logic is often in terms of dollars and cents rather than true value and sense. "I paid over a million dollars for this airplane, and I'm paying you a good salary to fly it. Now get me to my destination or I'll hire someone who can." It takes an independently wealthy pilot to resist such a threat. Also, a pilot who has just been angered by such a threat is not in the proper frame of mind to fly an airplane.

I told one student, "You're well-coordinated and you do a good job of physically flying the airplane, but unless you guard against your natural tendency to be impulsive, to act before you think, you're going to die in an airplane." I was wrong; he died of natural causes. But he always remembered what I told him and became a cautious pilot. Instructors have the responsibility not only of teaching people to fly, but also of warning them about personality traits that might prove deadly in the cockpit.

I have a tendency to be in a hurry. If I have a job to do, I like to get with it. I walk fast, drive fast, and much to my wife's dismay, I never *eat* a meal—I inhale it. However, knowing that this trait can kill me, I discipline myself to slow down when I approach an airplane with the idea of flying it. I take my time during a preflight. If it's a cross-country flight, the weather becomes of great concern. I think through every aspect of its potential. I know that I look at an airplane more closely than most pilots. Kicking the tires is not my method. I am more careful because it's my very nature to overlook important details.

I usually have many things on my mind. Because of this, I can become absentminded concerning the details that assure a safe flight. I know this, and as a result, I focus my attention upon the more important aspects of preflight and flying an airplane.

Flight safety starts with a proper attitude toward flying. We all have inherent personal deficiencies that need to be considered. However, just as we have learned to use a checklist, we can also discipline ourselves to turn off our tendencies toward self-destruction.

Many pilots have succumbed to the propaganda that emphasizes the safety of an airplane. For many years I have told my students that an airplane can kill them. The pilot who doesn't believe this is a target for destruction. Every student should start with a deep respect for an airplane. I emphasize the necessity of developing basic proficiencies, staying proficient, and exercising good judgment. Even the most ham-handed individuals can eventually learn to control an airplane. However, basic personality and the way it affects matters of judgment are far more difficult to deal with. A person's attitude toward life, sense of values, self-esteem, and concern for others are really the more important ingredients that determine whether that person will become a safe pilot.

The most effective flight instructors become skilled in judging human nature and watching for psychological traits that might be a threat to life.

Concerning the subject of this book, the aviation industry and the FAA have performed a gross injustice in minimizing flight training in stalls and spins. Yet, economics, politics, and selfish interests have portrayed flying as something anyone can do with a minimum of training. Nothing could be further from the truth. This very propaganda creates a false mental and psychological approach to flying and plants the seeds of false confidence. As a result, we continue to pay the highest of prices for our folly—that of human life.

I can not overemphasize the value of unusual attitude training. For this reason I strongly recommend a course in aerobatics along with spin exposure. As a rule, most aerobatic pilots are knowledgeable in spins. An airplane is a three-axis, all-attitude vehicle, and it is possible for it to be flown in any attitude accidentally or on purpose. No pilot is ever really proficient and competent who has not learned and retained basic aerobatic skills.

Although a theoretical knowledge of spins is helpful, there is nothing that can take the place of actual spin practice. The instinctive reactions resulting from sufficient practice are of primary importance.

There are very few spin "experts." Those who seem to be the most knowledgeable are the first to admit they have a lot to learn. It is a very complex subject. However, contemporary spin research has revealed a considerable amount of knowledge about why airplanes behave as they do during spins.

This book is written from a pilot's viewpoint. I have tried to be as sparing as possible with the numbers and complexities that put pilots to sleep.

Sammy Mason
Santa Paula, California

THANKS

I am fortunate to have such a large assortment of talented friends to call upon for help.

I would like to express my appreciation to Steve Kuehle for spending many hours reading through several drafts; John Margwarth for invaluable assistance; Tony LeVier, my former boss, for encouraging me to write about such a difficult subject; Wendell Dowling for his superb artwork; Jodi Ebersole for reading the manuscript to see if it was understandable from a student's viewpoint; Jody Auldridge for hours of proofreading; Annette Mason for typing the final manuscript; and Paul Stough, Jim Patton, Jim Bowman, Joe Chambers, Dan DiCarlo, and Phil Brown of NASA for their invaluable help.

Thank you, one and all!

S. M.

STALLS, SPINS, AND SAFETY

1

"Learning a little about flying is like leading a tiger by the tail — the end does not justify his means."
-P. T. Barnstorm

SPIN-TRAINING APOLOGETICS

The tail surfaces are all that are identifiable. The twisted and battered wings protrude from a blackened mass of melted aluminum. The engine and the mangled propeller have been shoved into the cockpit area by the authoritative hand of gravity.

The stall and spin and the resultant loss of life were inadvertent. The pilot of this airplane and the unwary friends aboard anticipated only the pleasure of flight, not the termination of their lives.

The word "inadvertent" is used frequently throughout this text. People do not invest in costly flight training and in even more expensive aircraft with the intention of shortening their longevity.

Only trained reactions and an educated basic intelligence can protect our fragile lives within the realm of flight. Pilots must not only be trained in the skillful manipulation of the controls but also know the deceptive areas of flight where they might become unwilling victims.

The essence of good basic flight training is to provide pilots with a protective armor of knowledge about—and dependable motor reactions to—potential hazards in their incipient stages.

A pilot need not be haunted by a lack of confidence and by uncertain reactions. With a properly trained pilot, flying is a reasonably safe mode of transportation. It is certainly the most pleasurable. However, as with any other form of transportation, it is only as safe as the individual at the controls.

AN UNSOLVED PROBLEM Unfortunately, the undesirable and often unexpected maneuver that killed many pioneer aviators is still surprising and killing pilots today.

Most modern airplanes are spinnable. This is a desirable characteristic for an aerobatic airplane, but definitely unacceptable for an airplane used for transportation.

Although a nonspinnable airplane is certainly within the state of the art, manufacturers are not altogether to blame for the lack of progress in developing safer designs. The stringencies of our times also affect the field of aviation. Basic design changes involve regulatory complexities and large sums of money. The reason we have not seen any dramatic changes in the design of general aviation aircraft is that such changes are uneconomical.

The knowledge and skill of the pilot should always match the airplane that is being flown. Because most airplanes are spinnable, spin training should be included in the preparation and licensing of pilots.

Some feel that more lives would be lost because spin training in itself is hazardous. This is true only because of the improper training of flight instructors. If all flight instructors were properly trained, spin training could be carried out with a high degree of safety.

The Federal Aviation Administration (FAA) seems to be reawakening to the need of spin training. Properly implemented, spin training would undoubtedly save many lives. Also, there is a moral aspect to be considered. It seems criminal to license a pilot who has never experienced a spin to carry passengers. Far too many pilots experience their first spin at an altitude from which they are too low to recover.

UNNECESSARY FEAR BREEDS INCOMPETENCE The fear of spins is understandable. We all fear the unknown. However, a good instructor can guide students through this initial fear and expose them to spins early in their flight training. It is senseless to continue flying with a fear that can be overcome in such a short time.

Pilots who have not been trained in spins and the recognition of incipient conditions from which inadvertent spins emerge develop grotesque flying habits. For example, they tend to fly their approaches with excessive speed, often overshooting runways and either colliding with obstructions at the end of the runway or attempting a last-minute go-around, necessitating a steep climb at low speed to clear obstructions. Many accidental spins have been the result of aborted landings that were prompted by panic and incompetence.

Some feel that all that is needed is training in the manner in which inadvertent spins develop so that pilots will learn to avoid those conditions. They argue that there is no need to expose pilots to spins if they know how to avoid them.

This approach to the problem reveals a misunderstanding of how inadvertent spins develop. Simply knowing how to avoid a spin may not prevent a pilot from entering one when distracted by other things.

During an air show in Denver, Colorado, a pilot competing in a balloon-bursting contest maneuvered his Fairchild 24 through wingovers, dives, and climbs in an effort to hit the air-filled toy balloon. The balloon descended lower and lower as it eluded the hapless pilot on each pass.

Finally, at a very low altitude, the fearless aviator swooped in from below and climbed steeply toward the balloon, determined to devour it with the propeller. The balloon was still out of reach when the airplane stalled. The airplane started to spin just as it hit the ground in a tangle of tubing, wood, and fabric. The balloon landed softly, the obvious victor. Fortunately, the pilot survived—an unusual occurrence after a stall/spin accident.

SPINS EVOLVING FROM ROUTINE FLIGHT SITUATIONS Although most stall/spin accidents are not the result of situations as dramatic as bursting balloons, they occur more frequently than they should, and from everyday flight situations common to normal aircraft utilization.

The answer is not more knowledge, although this is important. Avoiding accidental spins cannot be accomplished through knowledge alone, any more than thinking about placing one foot ahead of the other will improve your ability to walk. The pilot who must devote excessive thought to the manipulation of the controls and maintaining proper flight attitudes has not yet learned to fly well enough. Deciding to not get anywhere near the stall is not the answer. Flying at high angles of attack is part of normal aircraft utilization. Short-field operations, for example, require approaches at minimum speed.

While flying at minimum speed, we might be required to make a sudden pull-up, steep turn, or combination of both to avoid other traffic while on final approach. Or we might have to abort a landing well down the runway and pull up steeply at low speed to avoid hitting an obstacle. We may plan and hope that we never become involved in such situations, but flights do not always go according to plans. We must be able to fly an airplane through the complete range of its capabilities, and be able to do it instinctively.

REGULATORY BAND-AIDS Regulations, seminars, and biennial checks are the crutches of an inadequate approach to flight training. If pilots were adequately trained in the fundamentals of safe flying, many of these regulatory Band-Aids could be dispensed with.

The complexities of airspace, communications, navigation, and instrument flying are thrust upon pilots before they have mastered the fundamentals of aircraft control. As a result, many pilots go on to check out in complex aircraft and acquire a walletful of ratings before they have really learned how to fly.

Training in the conditions leading up to a spin and being able to recognize warnings that a spin is about to occur are important and should be a part of a pilot's stall/spin education, but these aspects are only part of the answer. The conditions leading up to the spin may be so completely camouflaged by other factors that demand attention that the pilot may not become aware of them.

Even experienced pilots have become stall/spin victims after complex emergency situations trapped them into hitting the ground in uncontrolled flight.

The answer is more thorough flight training, the type of training that gives one the ability to fly within the realm of a potential stall or spin instinctively and safely.

Incipient stalls and spins are common as aerobatic pilots practice their maneuvers. However, their recoveries are so instant and automatic that the incipient stalls and spins are scarcely recognizable. All the conditions may exist for a spin, but a touch of rudder and a flick of the wrist prevent even a small bobble in an attempted maneuver.

SPINS WITHOUT WARNING The aerodynamic buffeting that is experienced during stall practice will probably not exist during an inadvertent spin entry. During a ball-centered stall, most wings will stall at the root section first. Because the tail of the airplane is directly behind the root section of the wing, the pilot can feel this disturbance, which results when the turbulent air strikes the tail. However, when the ball is not centered and the nose of the airplane is yawed one way or another, a spanwise flow of air causes the wing to stall in a different manner and the initial stall occurs at or near the wing tip. When the stall is excited at one of the wing tips, a quick spin entry without the benefit of stall buffet warning may occur. There is nothing behind the wing tip for the disturbed air to buffet against. A stall from yawed flight usually results in at least the start of a spin.

As I was writing this portion of the book, I began to wonder whether it was possible to enter a spin without sounding the stall warning. I selected a Cessna Aerobat in which the stall warning was set to go off very close to the stall. Sure enough, after a little experimentation, I was able to enter a spin without actuating the warning. The stall warning port was located on the left wing. By spinning to the right I avoided actuating the warning.

It is evident that in the absence of buffet or audible warning or through unawareness of warnings because of distractions, the first indication of trouble may be the spin itself. How well the pilot flies the airplane at this moment may mean the difference between life and death.

The stall that results in instant-spin aerodynamics also produces instant-spin dynamics. In other words, the yaw and roll that are common to spins are present at the onset.

The more pilots are exposed to these dynamics and correct for them, the better they will be able to recognize and react to them. The full dynamics of a spin can only be experienced when the spin is fully developed.

EXERCISING MENTAL AND PHYSICAL SKILLS Flying involves mental and physical skills that must be exercised simultaneously. The flight instructor is the coach who should encourage the exercises that develop these abilities.

A winning game of tennis is the result of many hard practice sessions which develop the reactions, timing, and accuracy essential to victory.

Race drivers who have never experienced spinouts endanger themselves and other drivers on the track. Going into the first turn in a competitive pack of cars, they can be the catalyst that activates a massive pileup.

Drivers who have practiced recovering from fully developed spinouts on slick pans have developed the reflexes that cause them to react instantly and skillfully when they sense an incipient spinout while rounding a curve within inches of other cars.

Flying skills are best developed when we practice maneuvers that stretch our ability. Our goal should be a proficiency adequate to any situation that we might face while pilot in command. I recognize that on rare occasions pilots are faced with impossible situations which are beyond human skills. However, this isn't the reason for most accidents. Most accidents are preventable. They are usually the result of a pilot's inability to think and fly the airplane at the same time. Often, simple distractions result in serious accidents.

I recommend not only spin training, but also a course in basic aerobatics. An airplane is an all-attitude vehicle, and pilots should be able to recover from any attitude. The elimination of unnecessary fears and the confidence gained will add to the pleasure of flying.

As I gathered material for this book, I talked to many pilots concerning the pros and cons of spin training. Most of the pilots did not have spin training; however, the majority felt that they should have been trained in spins.

I asked those who had spin training to record as best they could their initial feelings about being exposed to spins. I also asked them to comment on the value of receiving such training. I have included excerpts from several of them.

. . . my instructor treated them as just another maneuver to be learned. It was not tentative, but another step in the learning process. I wasn't afraid because his confident attitude assured me that there was no unusual hazard involved. I feel that the experience of spins and other unusual attitudes has increased my confidence and made me a better pilot.—*Gene Evans, actor and private pilot*

My instructor believed that students should have spins before solo. I experienced them early in my flight training. I was confident in my instructor's ability and did not give them a second thought.—*Joe Murray, geologist and private pilot*

. . . because of a qualified instructor, I felt very comfortable in all flight attitudes. My first experience in spins came during my third lesson. The first one was a little frightening and disorienting, but after that they didn't bother me. . . . I think knowledge, experience, and awareness of spins and other unusual attitudes are important to all pilots.—*Jim Stallings, businessman and private pilot*

My first experience in spins was when I was a student pilot practicing stalls solo. I was able to recover, but needless to say, I was scared!

After I received my private license, I found a competent instructor and learned spins and other basic aerobatics. I am no longer afraid of spins, and I feel confident that I can recover from any attitude. I strongly recommend aerobatic training.
—*Judy Higby, private pilot*

I am distressed by the fact that the FAA does not require spin recovery proficiency for a private license. . . . spin recovery exercises were mandatory during my flight instruction. These exercises were preceded by ground school lessons explaining the aerodynamic principles affecting entry and recovery from spins. The first spin, of course, produced a unique visceral experience. Through successive spins and training in basic aerobatics, these maneuvers became enjoyable.—*Eloy Molina, attorney*

Looking back in my log, I had spins at three hours and thirty-five minutes. After solo, it was spins, stalls, and landings, repeatedly. I remember apprehension, thrill, some nervousness, coupled with the knowledge that I was being introduced to a regime totally new and unknown, and that it was a thrilling challenge. It took a while for the apprehension to die down, even after I was practicing solo spins. Most of my solo flights included spins, so I guess they got to be fun.

As to my personal attitude towards them in training, I think we are crazy in not having the student feel thoroughly confident in the maneuver, by having shown him and having him practice them solo.—*Bill Mason, commercial pilot*

START SMART

Immediately reject all thoughts of attempting to develop spin proficiency without the benefit of an instructor who is qualified in spins! Many years ago, before spin requirements were dropped as part of the licensing requirements for pilots, all flight instructors were capable of teaching spins. However, not many instructors from the old school are still in business and not many of today's instructors know enough about spins to teach them properly.

It's interesting that the FAA has made it legally unsafe for an instructor to teach spins. Those seeking an instructor's rating need only show through a logbook entry that they have performed a one-turn spin in each direction. This entry must be certified by an instructor who also may not know anything about spins. Most airplanes are still in the incipient stage during the first turn of spin rotation, so even an instructor who has performed the spins essential for an instructor's rating may have never actually demonstrated the ability to properly recover from a fully developed spin.

If you do not know of an instructor who is competent in this realm, seek out one of the better aerobatic schools that are scattered throughout the United States. Take your time. Interview the school management and the instructor who will be teaching you.

CHOOSING A SCHOOL

I have found that the schools that take pride in their equipment and training facilities usually do a better job of flight training as well. Do the personnel present a good appearance? Are the training facilities neat and clean? Does the training aircraft appear clean and well maintained? Of course, this doesn't always tell a true story, but very frankly, I would walk away from a school that didn't present a good appearance. I have found that one item of appearance that usually coincides with an airplane's true condition is the state of an aircraft's belly. If the belly is caked with oil and dirt, the rest of the airplane is probably not very well maintained. See Figure 1.1.

It is wise not to sign up for a course or to purchase a block of flying time until you have made a trial flight with your instructor. In the final analysis, it is the flight instructor who will be the key to how much you will learn and your rate of progress. If you're not satisfied with your flight instructor, either request another one or seek out another school. A good flight instructor will lead you through a course on spins, or through a complete aerobatic course, with a minimum of apprehension. If the instructor is more interested in demonstrating aerobatic skills than doing a competent job of teaching, you're obviously not going to get your money's worth. Beware of the instructor who wrings you out during the first lesson. A good spin or aerobatic lesson will leave you more confident than before. Part of an instructor's job is to instill confidence in the student.

FIGURE 1.1 Choose a school that takes pride in its operation.

THE HAZARDS The first step in flying safely is to have a thorough understanding of the hazards involved. You should know from the outset that most conventional airplanes are capable of spinning under the right circumstances. You should also know that *most* airplanes are not approved for intentional spins. The placards that read "Intentional spins prohibited" mean one of two things: either the airplane has not been spin-tested, or its spin characteristics have been observed only in their very early stages of development.

Some airplanes are approved to be spun in the utility category only, which simply means that there must not be anything in the baggage compartment, or anyone occupying the rear seats. In other words, it can only be spun at a forward center of gravity (c.g.), where it will be very spin-resistant. Generally, you will not obtain a true picture of what a real spin is like under these conditions. An inadvertent spin is most likely to occur when the airplane is heavily loaded and near the aft c.g. limits. If this airplane were to be spun at an aft c.g., it would very likely display either poor, or unrecoverable, spin characteristics. It is important that airplanes that are approved for spins be configured as recommended by the manufacturer. In most cases, this means that the spin center of gravity limits be strictly adhered to.

Never spin a normal category airplane! Some pilots, knowing that the manufacturer has had to demonstrate a one-turn spin with a recovery performed within one additional turn, feel that as long as they don't exceed one turn, it is safe to do spins in a normal category airplane. Nothing could be further from the truth.

Let's look at the FAA requirements for spin-testing normal category aircraft.

Spinning

(a) Normal category: A single engine, normal category airplane must be able to recover from a one-turn spin or a three-second spin, whichever takes longer, in not more than one additional turn, with the controls used in the manner normally used for a recovery.

In addition—

(1) For both flaps retracted and flaps extended conditions, the applicable airspeed limit and positive limit maneuvering load factor may not be exceeded;

(2) There may be no excessive back pressure during the spin recovery; and

(3) It must be impossible to obtain uncontrollable spins with any use of the controls. For the flaps extended condition, the flaps may be retracted during the recovery.[1]

Now let's really look at these requirements. First, a test pilot has time to practice the "normally used" recovery technique—whatever that may be. Although the full range of c.g. and weight conditions are investigated, as well as the various flap and gear configurations, the tests are performed under the most ideal conditions and the demonstration pilot, knowing fully what to expect from previous exposures, is alert and ready.

Second, any airplane that rotates one additional turn after the application of recovery controls following only one turn of spin has demonstrated *horrible* recovery characteristics! Yet it has met the requirement for a normal category airplane. Remember, normal category doesn't mean normal spin characteristics. It is very possible that if the pilot were to permit the airplane to continue rotating beyond one turn of spin, the airplane could become unrecoverable!

Perform spins only in airplanes of the utility or aerobatic categories. However, not all utility category airplanes are approved for spins. Only those *approved* for spins should be spun. Airplanes approved for spinning must comply with the following requirements:

(c) Acrobatic category: . . .

(1) The airplane must recover from any point in a spin, in not more than one and one-half additional turns after normal recovery application of controls.

[1]FAR (Federal Air Regulation) 23.221.

Prior to normal recovery application of the controls, the spin test must proceed for six turns or three seconds, whichever takes longer, with flaps retracted, and one turn or three seconds, whichever takes longer, with flaps extended. However, beyond three seconds, the spin may be discontinued when spiral characteristics appear with flaps retracted.

(2) For both the flaps-retracted and flaps extended conditions, the applicable airspeed limit and positive limit maneuvering load factor may not be exceeded. For the flaps-extended condition, the flaps may be retracted during recovery, if a placard is installed prohibiting intentional spins with flaps extended.

(3) It must be impossible to obtain uncontrollable spins with any use of the controls.[2]

It is obvious that airplanes which are approved for intentional spinning are safe to spin. Unfortunately, as pilots gain proficiency and confidence by spinning airplanes that are approved for spins, they are sometimes tempted to toy with unapproved airplanes. This could easily result in a premature planting in a marble orchard.

From a training standpoint, some of the older general aviation lightplanes are best for spins. They will drop into a spin easily, but will recover just as easily.

The Cubs, Taylorcrafts, Aeronca Champions, Luscombes, Interstates, Porterfields, Rearwins, and Cessna 120/140s are all excellent spin trainers. The Citabria retains the same basic design as the "Champ" and is a good spin trainer. However, the Decathlon, which, except for a different wing, is nearly identical with the Citabria, is undoubtedly the best all-around unusual-attitude trainer. It enters and recovers from spins easily and, with an inverted fuel-and-oil system, is ideal for inverted spins as well.

Some of the old World War II primary trainers were also excellent spin trainers. The Stearman PT-17, Ryan PT-22, and the Fairchild PT-19 had good entry and recovery characteristics.

Although spins place very little stress upon airplane structures, these older machines should be thoroughly inspected to see that the termites are still holding hands.

If you prefer a more modern version of two-winged adventure, the Pitts S2-A is also a good spin trainer.

The Cessna 150 series, perhaps the most available of training aircraft, is a good spin trainer if the instructor knows how to overcome its spin resistance and avoid a half-spin, half-spiral type of gyration.

[2] FAR 23.221.

Some 150s are more spin resistant than others. Some can be spun quite easily, while others require a bit of trickery to make a positive entry and continue into a fully developed spin. The 150 has enough of the old fashioned traits to make it a good spin trainer. All of the yawing, cross-control situations that result in inadvertent spins can be effectively demonstrated in the Cessna 150. (Cessna publishes an excellent booklet on the spin characteristics of their 150 through 177 models. It can be purchased at Cessna training centers.)

Most spin-resistant airplanes will spin quite easily if a small amount of power is used during the entry to provide better airflow over the elevators and rudder. Also, most airplanes will make a cleaner entry if the entry is forced 2 or 3 knots over the normal 1-*g* stall speed. The tail surfaces are flying in less disturbed air at the time and are more effective. If power is used during the entry, the throttle should be closed as soon as autorotation takes place.

With a competent instructor, it is safe to spin any airplane that is approved for spinning. Nevertheless, it is important to have the airplane configured properly and to adhere to the c.g. limitations.

SUMMARY

1. Inadvertent spins are the result of a lack of familiarity and training in regard to spins and spin-inducing situations.
2. Trained reactions and education in the hazards involved are imperative.
3. Good flight training provides both knowledge and motor reactions against hazards in their incipient stages.
4. Spin and aerobatic training will improve your skills and provide confidence.
5. Pilots' knowledge and skills should always be a match for the airplanes they are flying.
6. The fear of spins can be eliminated through proper exposure.
7. A lack of stall/spin proficiency results in grotesque flying habits.
8. Distractions, combined with a lack of automatic flying skills, can result in inadvertent spins.
9. The proper utilization of an airplane requires that it frequently be flown at a high angle of attack.
10. The conditions leading up to a stall or spin can be camouflaged by other factors.
11. The first indication of loss of control may be the spin itself.

12. The most skilled reactions are developed while experiencing the full dynamics of a spin. They are best developed from flight exercises that stretch your ability.
13. Most flight instructors are *not* qualified to teach spins.
14. Seek out a qualified instructor. An aerobatic school is a good place to start.
15. Spin only airplanes that are approved for spinning.
16. An airplane that is approved for spins in the utility category might be dangerous to spin in the normal category because of the weight and c.g. difference.

"The Law of Gravity is not a general rule."

-P. T. Barnstorm

LIFT, STALL, AND MUSH

At an air show where I was performing, I watched with fascination as the local pilots engaged in a spot landing contest. As they matched their skills with one another, they tried to see who could come closest to a designated spot on the runway. Several contestants had tried, but none had come close.

Then, a Cessna 140, flown by an indomitable aviator, swung into the landing pattern. He was quite high as he turned onto final. To lose altitude, he pulled the nose up until he was near the stall and then began mushing until he had a near-parachute-descent angle. He was still high, however, as he approached his target. The closer he came to the spot, and the ground, the more he pulled the nose up. By now, he was falling out of the sky like a rejected demon. I thought surely that a measure of power was in order, but as he fell to the ground, the elevators seemed deflected fully upward as the propeller swung lazily at idle rpm.

Contact with the ground was sudden and dramatic. The landing gear spraddled grotesquely. Unable to suppress its gravitational thrashing, one of the gears broke loose and tumbled away from the impact. The windshield popped out of its frame and landed on the target—the only part of the airplane to make it.

The pilot and his passenger seemed depressingly dazed as they semi-waltzed away from their pitiful pile of aluminum. It was a shame to offer such a sacrifice and still fall short of the spot!

ANTI-SPIN REFLEXES The FAA claims that many stall/spin accidents are the result of distractions. The accident described above may be proof of that. The pilot had what the military calls "target fixation." His total attention was given to hitting the target. However, it is my opinion that the problem goes deeper than that. Developing safe flying skills involves the development of deeply ingrained reflexes that respond correctly to every condition of flight. A pilot who is exerting a conscious effort to manipulating the controls and maintaining the correct flight attitude has not yet reached the level of basic proficiency that is essential to safety.

THE TREACHEROUS MUSH

Many accidents that fall under the category of stall/spin do not involve a spin and are not full stalls. They are, like the above-mentioned crash, the result of "mushing," that is, being almost or partially stalled. Because the airplane didn't "feel" stalled, the pilot felt secure. The extreme descent rate may go unnoticed until the airplane is in proximity to the ground. Then, too often the first reaction is to further increase the angle of attack instead of applying power. The moment the pilot pulls back on the stick/wheel, the airplane stalls. Whether or not the airplane spins at this point depends upon the characteristics of the airplane.

The mush can be just as deadly as the stall or spin. An airplane with a heavy wing loading can fall out of the sky like a waxed anvil.

Mushing is that condition of flight in which induced drag is high due to a high angle of attack, but in which the angle is not quite high enough to induce a stall. Mushing occurs under the condition often referred to as the "backside of the power curve," where the induced drag is so great that the power required to keep the airplane flying is far greater than at low angles of attack.

We usually think of mushing as something that occurs at low speed, but like a stall, it can occur at any airspeed. I witnessed a good example of a high-speed mush at Paris, France.

I had just touched down in the Lockheed 286 Rigid Rotor helicopter and was pushing the collective down when the French aerobatic team lifted off the runway in their agile Fouga Magisters. It was the last day of the 1967 International Paris Air Show. I had flown the next to the last act, and the French team was to perform the grand finale. It was beginning to look as if we were going to make it all the way through the 5-day show without a fatality.

My interpreter, one of the members of the French team, greeted me as I climbed from the helicopter. We sat down on the grass to watch the performance. I marveled at the French team's precision as they rolled and looped in tight formation. Then, as they looped to the inverted position trailing colored smoke, they entered a dive, and each Magister separated from the formation in a "bomb-burst" maneuver. They seemed awfully low as the pilots attempted to recover from the dive. One of them, heading directly toward the crowd, mushed through the recovery arc and hit the ground. A ball of sickening orange flame rolled with the wreckage to within 100 ft of the crowd. Black smoke tumbled upward in a ghastly reminder that gravity often has the last word.

The most insidious and deadly form of mushing occurs during the curved flight that is experienced when pilots attempt to flare from a steep approach, make a tight turn within a narrow valley, pull from a dive at low altitude after a spin recovery, or break out of a low overcast during a steep descent angle. Curved flight can be in the form of a turn, any attitude within the circle of a loop, or both.

Curved flight produces an effect on stall speed which is similar to that of weight. In other words, if you increase an airplane's weight, you also increase its stall speed.

There is a point prior to the stall where an airplane descends, and even though you may increase the angle of attack, you can't stop it. The reason is that the drag bleeds off the airspeed so rapidly that you can't stop the descent before the airplane stalls. Nearly all pilots have been in landing situations in which they were unable to flare completely from a steep descent angle. You may have been settling rapidly toward the runway, and when you went to rotate to a higher angle of attack, the airplane stalled and hit the runway hard. Although stalls occur because of excessive angle of attack, it takes airspeed to sustain the increased weight resulting from more *g*'s. If the speed diminishes, the *g*'s (or portion thereof) required to negotiate the curve cannot be maintained.

During a 1-*g* stall, a pilot can see the nose attitude relative to the horizon and knows that the wing must be at a high angle of attack and near the stall. However, during a turn or other curved flight maneuvers, the only clue a pilot may have (other than an angle-of-attack device) that a stall or mush is imminent is pressure against the seat of the pants, in other words, the amount of gravity that can be felt or interpreted through an accelerometer. When maneuvering through critical curved flight situations, the pilot needs the feel and finesse of someone milking a mouse.

During basic flight training there is not enough emphasis on stalls that are entered from unusual attitudes. One reason that I emphasize aerobatic training is that you will experience stalls from every conceivable attitude of flight. The most deadly stalls occur from attitudes that a pilot has never experienced. For example, students should be able to recognize and recover from stalls that are entered from a steep nose-down attitude. Many inadvertent stalls and spins have occurred from just such an attitude.

MAXIMUM LIFT COEFFICIENT The amount of lift that a wing can produce is always limited. The term "$C_{L\ max}$" may perplex those unfamiliar with aerodynamic palaver, but if the wing could talk to the pilot at the moment of developing its maximum lift coefficient, it might say,"That's the best I can do, dunderhead! Tilt me any higher and I'll dump your whole load."

ANGLE OF ATTACK Every airfoil shape has an angle of attack at which it develops its maximum lift. When that angle is exceeded, the lifting process falls apart and the wing stalls. See Figure 2.1.

As the wing develops lift, if the angle of attack is increased, the low pressure on top of the wing increases. This further develops a positive pressure gradient, resulting in an increasing mass airflow from the high pressure ahead of the wing to the low-pressure air above the wing. The low pressure attracts the high-pressure molecules of air. They rush toward the leading edge of the wing, swoop over the top of the airfoil at a speed that exceeds the airspeed of the airplane, and fling themselves downward aft of the wing, creating lift.

When the $C_{L\ max}$ angle is exceeded, the lifting process breaks down. The center of pressure, now forward, begins to draw high-pressure molecules onto the upper rear part of the airfoil. The molecules that until now had been moving along the airfoil contour from the leading edge begin to collide with a heavy population of molecules near the trailing edge. When this happens, they become unstable, in a negative pressure gradient, and separate from the airfoil contour. This is the initial phase of a stall. As the angle of attack is further increased, the stall progresses toward the leading edge until, finally, the entire wing is stalled. It should be noted here that stalls can start at the leading edge and work aft. This is especially true of wings with sharp leading edges, or where a wedged stall strip has been installed.

WING CONFIGURATIONS Wings come in various shapes, sizes, and aerodynamic personalities. Just about every shape has been tried, and most of them have flown. Some are fashioned for flying high and fast, some for flying slow and taking off and landing in short distances, and some for powerless soaring. Some are even designed to work in a helical path and provide combined lift and thrust for helicopters. And, yes, given enough power, even a barn door will fly. However, when it was on the barn it was probably near the stall and it wouldn't do any better aerodynamically!

The overall shape of a wing, including its airfoil and planform, will strongly influence its stall characteristics. See Figures 2.2 to 2.8.

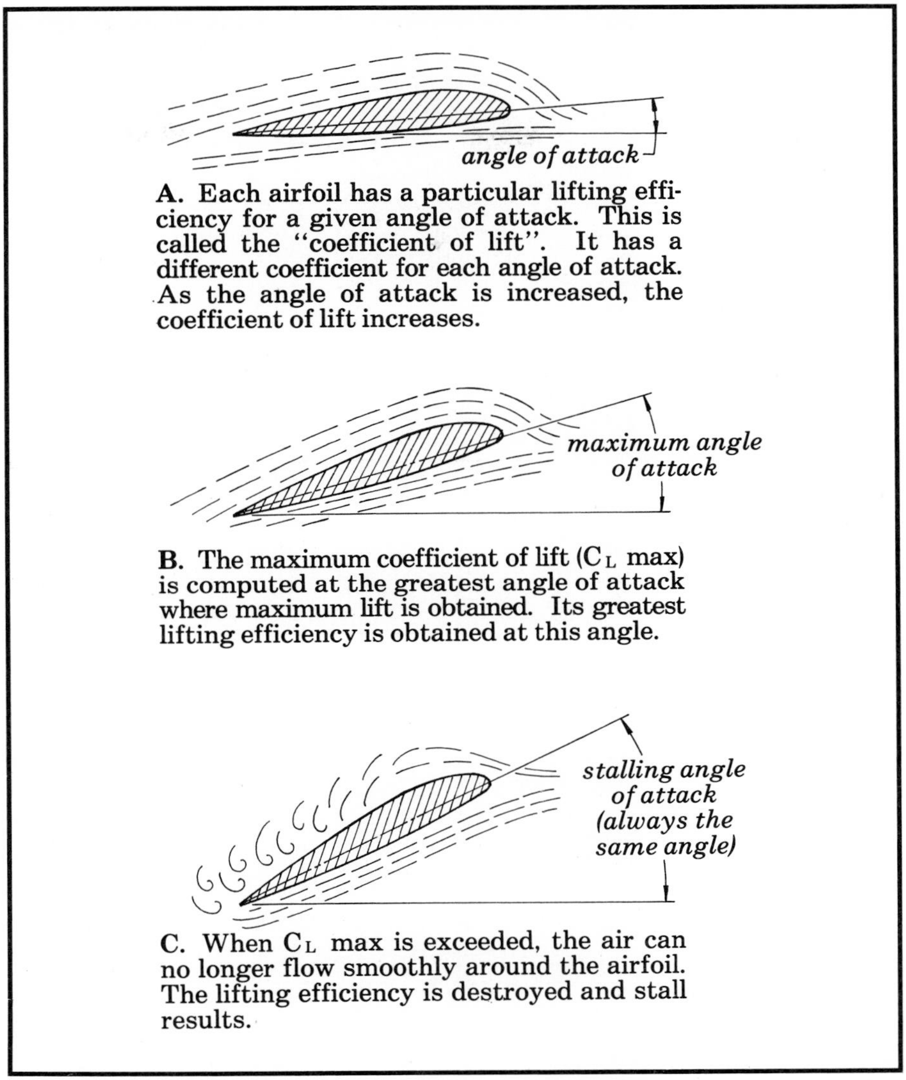

angle of attack

A. Each airfoil has a particular lifting efficiency for a given angle of attack. This is called the "coefficient of lift". It has a different coefficient for each angle of attack. As the angle of attack is increased, the coefficient of lift increases.

maximum angle of attack

B. The maximum coefficient of lift (C_L max) is computed at the greatest angle of attack where maximum lift is obtained. Its greatest lifting efficiency is obtained at this angle.

stalling angle of attack (always the same angle)

C. When C_L max is exceeded, the air can no longer flow smoothly around the airfoil. The lifting efficiency is destroyed and stall results.

FIGURE 2.1 **(*a*) Angle of attack, (*b*) maximum coefficient of lift, and (*c*) the stall.**

Most planforms tend to stall first in one part of the wing, usually either at the wing tip or the root section. Only the elliptically shaped wing tends to stall simultaneously over the entire span. This is because this planform has constant local lift coefficients throughout the span. With a constant lift distribution, all portions of the wing will reach a stall at the same angle of attack.

FIGURE 2.2 Oil flow patterns on Grumman Yankee wing at zero angle of attack. *(NASA.)*

FIGURE 2.3 Oil flow patterns on Grumman Yankee wing at 12° angle of attack. *(NASA.)*

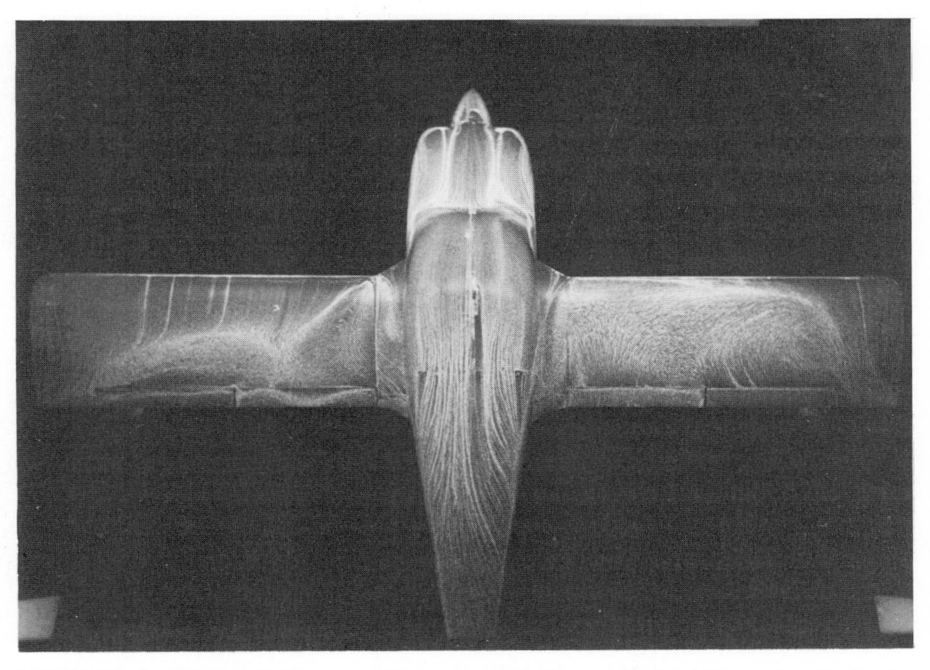

FIGURE 2.4 Oil flow patterns on Grumman Yankee wing at 14° angle of attack. *(NASA.)*

FIGURE 2.5 Oil flow patterns on Grumman Yankee wing at 14° angle of attack showing reversal of stall patterns. *(NASA.)*

FIGURE 2.6 Oil flow patterns on baseline of Grumman Yankee wing at 20° angle of attack. *(NASA.)*

FIGURE 2.7 Oil flow patterns on outboard droop configuration of Grumman Yankee wing at 20° angle of attack. *(NASA.)*

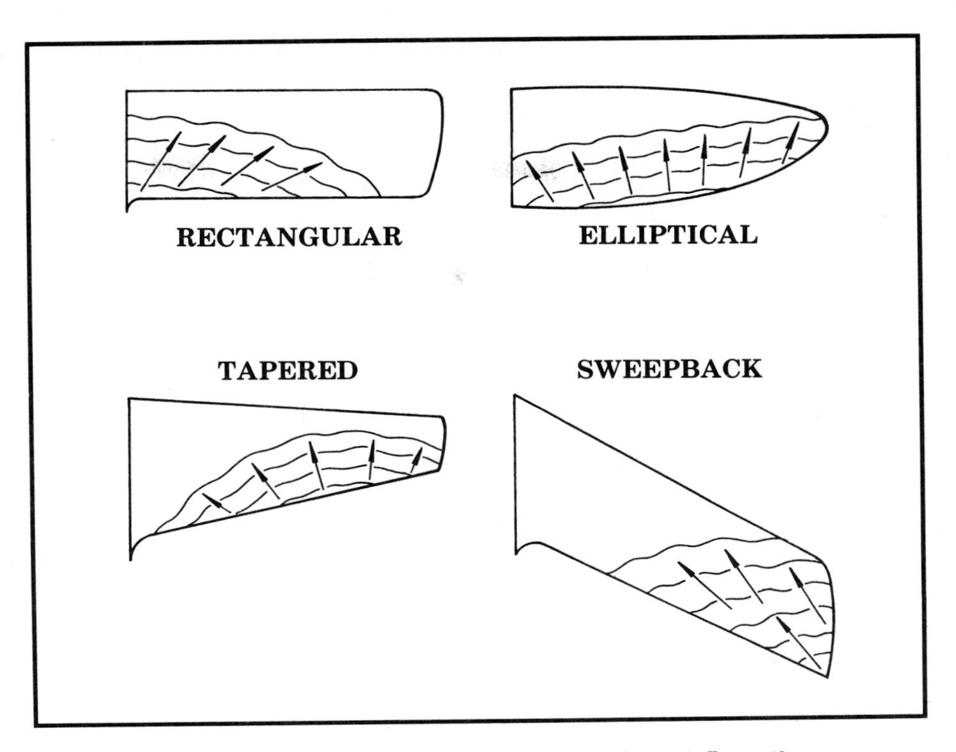

FIGURE 2.8 Stall progression patterns of various wing configurations.

Compared with other planforms, the rectangular-shaped wing, a rather inefficient wing, possesses the best stall characteristics. It exhibits low lift coefficients at the tip and high lift coefficients at the root. Because stalls are initiated in the areas of the highest angles of attack, the rectangular wing will stall in the root section first. This stall pattern is desirable because it provides adequate stall warning in the form of aerodynamic buffet while the ailerons, still flying in undisturbed air, provide good lateral control. The buffeting that the pilot feels is the result of the turbulent air from the stall, tumbling against the fuselage and the horizontal stabilizer.

The degree of aerodynamic buffet warning will vary with the airplane. Large aircraft sometimes possess very strong initial stall buffet. With the crew sitting a considerable distance forward of the wing, the buffeting becomes so amplified that it falls into the violent category. My experience with the Lockheed Constellation was that you'd better have your seat belt tight during the stall if you wanted to stay seated. The airplane had excellent stall characteristics, but the buffeting was of very high amplitude. The pilot had little doubt that he was entering a stall.

Although the characteristics existing in rectangular-shaped wings are not inherent in other planforms, aerodynamicists attempt to alter the wings in various ways to ensure that the initial phase of the stall occurs in the root section first.

PITCH DUE TO STALL On a normally stable airplane, with the center of gravity within limits, the horizontal stabilizer will always be producing negative lift to counteract the basic tendency of the wing to pitch downward around the c.g. Except in T-tail configurations, the downwash from the wing as it strikes the horizontal stabilizer is often a significant factor in producing this negative lift. When the root section of the wing stalls, this downflow against the stabilizer is destroyed and the nose pitches downward. This characteristic is desirable because as the nose pitches down, it tends to reduce the angle of attack and restore a smooth airflow around the wing. However, if the stall occurs outboard of the influence of the horizontal stabilizer, this pitch-down reaction may not occur. In fact, a pitch-up reaction may take place, which further increases the angle of attack and deepens the stall. See Figure 2.9.

The tapered wing planform tends to stall near the wing tip first. This results in a lack of buffet warning and a premature loss of lateral control. The use of ailerons to level the wings under these conditions only makes matters worse. The depressed aileron on the stalled wing aggravates the stall and results in additional drag that tends to yaw the airplane into a spin.

When I was performing with the Hollywood Hawks air show, a fellow performer, Roy Cusick, built a very small high-speed glider with a wingspan of only 14 ft. The wings were from the Schoenfeldt Firecracker, a racing airplane made famous by Tony LeVier. It was a highly tapered thin wing, and Roy had cut 4½ ft out of the center section, and then had joined the remaining lengths together with a common fitting. It was tremendously fast for a glider. I remember seeing 225 miles per hour indicated in a not-too-steep dive.

After making a few test flights and developing an aerobatic routine, Roy was ready to fly his first air show at San Diego, California.

I towed him to 2500 ft behind my 450-hp Stearman biplane. Roy then jettisoned the towline and started his act. I knew his routine, and when he was supposed to perform a loop, the tiny glider entered a series of extremely rapid snap rolls. Thinking little more than that Roy had decided to change his routine and that the snap rolls were indeed spectacular, I dismissed it from my mind. However, when I talked to him on the ground he said, "I'm going to have to modify the wing. When I tried to do a loop, it did a bunch of snap rolls instead."

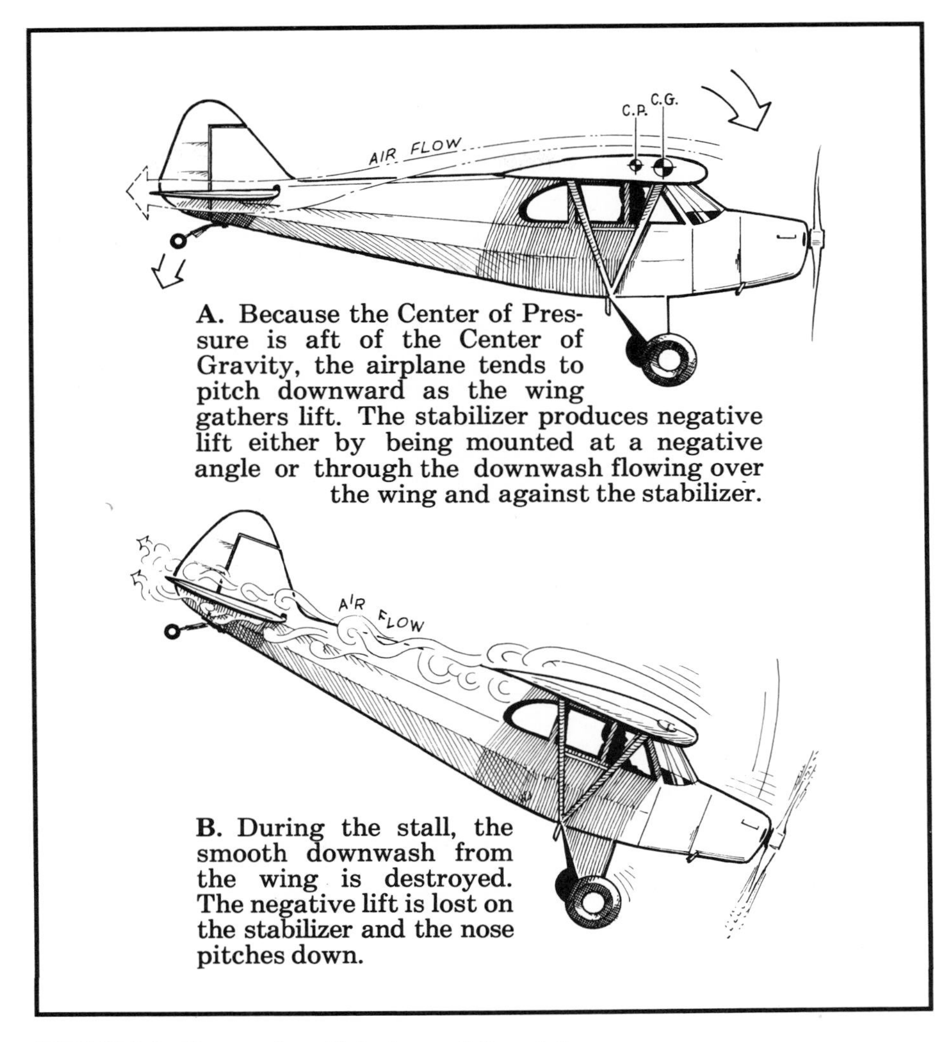

FIGURE 2.9 Reason for pitch-down at the stall.

His modification consisted of recontouring the wing tips so that they were essentially "washed out," or had less angle of attack than the rest of the wing during the stall. This permitted the inboard section of the wing to stall before the wing tips, which resulted in a more controllable and predictable stall.

TAILORING GOOD STALL CHARACTERISTICS There are several methods used to control the stall pattern of a wing. See Figure 2.10. One procedure is to vary the

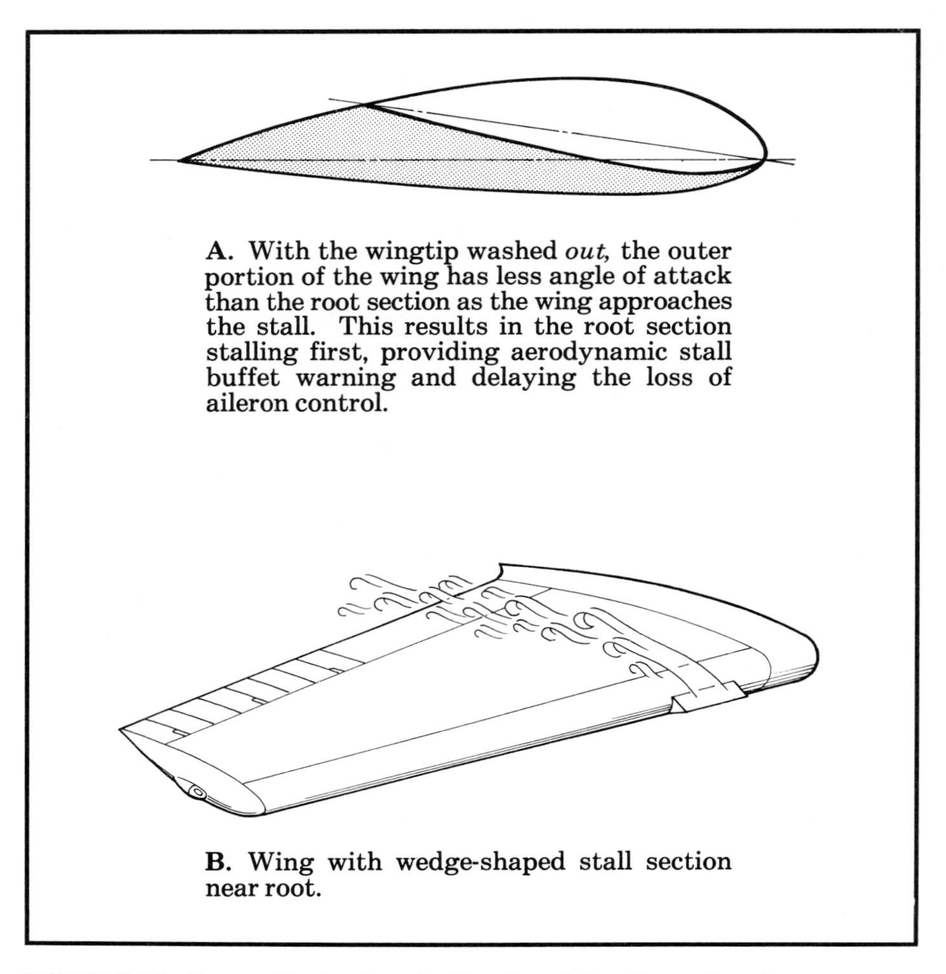

FIGURE 2.10 Two methods of controlling the stall pattern of a wing.

airfoil shape throughout the span to provide a variation of local stall angle of attack; the idea would be to provide high lift coefficients in the vicinity of the root section so that it will stall first. Near the wing tips the airfoils are contoured to induce lower local lift coefficients as the root section stalls.

Another method is to use a uniform airfoil section throughout the span, but to twist (wash out) the wing so that the outboard section of the wing is at a reduced angle of attack as the root section stalls. This method is easily accomplished on fabric-covered wings, where adjustments are provided at the strut attachment fittings. On metal wings, the washout process is accomplished during manufacture, while the wing is in a jig.

Thin airfoil shapes provide lower lift coefficients than thick ones. This effect can also be produced by installing a wedge-shaped strip along the leading edge of the portion of the wing that should stall first. These wedge-shaped strips can frequently be seen near the root sections of wings. The effect is to stall the root section before the wing tips.

Sweptwing configurations without modifications such as slats, leading-edge flaps, and wing fences possess rather severe tip stall characteristics. The outboard sections of a swept wing trail the inboard sections. The outboard low-pressure areas tend to draw the boundary layers toward the tips. This results in thickening the boundary at the tips, which easily separates at high angles of attack, inducing a stall.

Because the center of pressure of the wing moves well forward when a swept wing stalls, it creates the additional problem of the airplane pitching up.

It is interesting to note that when pilots become sloppy in their rudder coordination and permit adverse yaw to develop, the same conditions that prevail in a swept wing can occur in any other wing configuration. For example, even the docile rectangular wing can become hostile when sloppy flying permits a cross flow across the wing, resulting in boundary layer separation and tip stall.

Many aerobatic pilots prefer sweptwing airplanes because of their rapid spin and snap-roll characteristics. Another reason for their rapid snap-roll and spin rate is that the lifting wing produces even more lift as it swings forward. As the retreating wing, already stalled, swings aft, it produces less.

FACTORS THAT DESTROY LIFT

Pilots should be aware that any deformation of the original wing configuration can greatly reduce the amount of lift it produces and degrade the stall characteristics. During my air show days, I had an experience with a Curtiss JN-4D "Jenny," which dramatized the effect of changing the original shape of the airfoil.

After an air show in Chadron, Nebraska, I decided to mount a flat tank with square, sharp edges on top of the center section area of the Jenny. The Jenny's only claim to efficiency was that it could consume more fuel while going nowhere than any other airplane I had ever flown. My idea was to increase the fuel capacity so that we could fly longer legs. I knew that the tank would destroy the lift over the center section, but I figured that the Jenny had more than enough wing area and that destroying a little of it for the sake of more fuel capacity was worthwhile.

When I had completed the installation, I hopped the Jenny around the field to see if its performance had changed enough to invoke concern. It seemed to lift off quickly, even at Chadron's 3300-ft elevation, and I noticed very little deterioration in performance.

However, the next day, while on a westward cross-country flight, I pumped a little adrenalin during a takeoff from the Rock Springs, Wyoming, airport. The 6700-ft elevation made the difference. I couldn't get the Jenny out of ground effect. The airport was situated on a bluff, and when I went off the end of the runway, the Jenny settled below the airport elevation. I watched the engine temperatures rise as I attempted to climb high enough to get the airplane back on the runway. Finally, by flying back and forth near the windward side of the bluff, I picked up enough ascending air to do the trick.

After landing, I removed the tank from the top of the wing. The next takeoff was almost spectacular compared with the previous one.

Never again did I mount anything on the upper surface of an airfoil. Objects can be mounted on the bottom of a wing without greatly affecting lift, but upper surfaces of a wing should be kept aerodynamically clean.

Deformations of the leading edge of a wing are especially conducive to destroying smooth airflow over the upper surface and contribute to poor stall behavior. Ice accumulations can alter the shape of an airfoil. The change in shape can reduce lift and increase stall speed and drag. Airplanes with normally docile stall characteristics can become violent when the wings are deformed by ice. An ice-laden aircraft, in combination with a difficult instrument approach into low ceilings and reduced visibility, presents hazards that have reduced the pilot population on more than one occasion.

Also, a thin layer of innocuous-appearing frost can so greatly retard the boundary layer flow over a wing that a premature air separation can take place and a stall can occur. Frost is a rough surface with many tiny sharp peaks. These peaks disturb the air flowing over the top surface of a wing, resulting in boundary layer separation and a premature stall. Obviously, small aircraft with minimal power will be more affected than very large aircraft with plenty of thrust. However, the performance of *all* aircraft is reduced by a layer of frost. Even large aircraft have been known to run the length of a runway and crash during critical takeoff situations, all because of a thin layer of frost which reduced lift while it increased drag.

Although the wing and tail surfaces are the most critical surfaces, frost, ice, and snow should be removed from the fuselage as well. The fuselage contributes significantly to an airplane's lift. If icing conditions prevail during or after takeoff, frost provides a nucleus for the buildup. In other words, ice can more easily adhere to the airframe.

Ice formations can also change the mass balance of wings and control surfaces, leading to destructive aerodynamic flutter. Water leaking inside of the wing can run to the trailing edge, where it accumulates. As the temperature drops, it freezes. A friend of mine nearly shucked the wings from his Cessna 182 as the entire wing started to flutter when he encountered turbulence. Upon landing, he observed water dripping from the trailing edge. Closer observation revealed a large accumulation of ice in the aft portion of the wing. Ice can form inside a wing as well as outside and can be just as hazardous!

Cold weather preflights are by nature most miserable. No one enjoys checking over an airplane in a cold wind with the temperature below zero. However, if there ever is a time when a preflight is important, it is during cold weather. Cold weather preflights take twice as long, but they are also twice as important. Hasty pilots who refuse to take the time to clear the surfaces of ice, snow, or frost may suddenly run out of the time they are so desperate to save.

Research is beginning to disclose that heavy rain can greatly decrease the lifting potential of a wing. It is now thought that many of the accidents that were formerly attributed to wind shear may have been caused by the destruction of lift resulting from encountering heavy rain. It has been estimated that the effect of a roughened airfoil can reduce lift by 30 percent at the angles of attack that characterize the approach or missed-approach profile. The increase in drag on both the wings and fuselage is estimated to be on the order of 5 to 20 percent, with penalties up to 30 percent possible.

Unlike wind shear, heavy rain can be seen, and should be avoided wherever possible, especially during takeoffs and approaches.

NO SPEED—NO STALL

A maneuver included in a basic aerobatic course is the "hammerhead stall." This is a misnomer because, when the maneuver is done properly, the airplane never stalls, even though the airspeed may drop well below its normal 1-*g* stalling speed. This maneuver is a practical way to demonstrate that stalls are always the result of an excessive angle of attack and not a lack of airspeed.

The maneuver is performed by gaining enough airspeed to zoom the airplane into a vertical climb. Then, just prior to losing all forward speed, full rudder is applied to pivot the airplane about its yaw axis and into a vertical dive. In this attitude, speed is gained rapidly, and the pilot recovers from the dive.

While going straight up, just prior to the pivot maneuver, the back pressure that was being held against the stick to loop to the vertical position is released, and the angle of attack becomes zero. Even though the airspeed drops well below the normal 1-g stall speed, the airplane will not, indeed cannot, stall because the wing has no angle of attack.

This is just one of the beneficial maneuvers learned in an aerobatic course. However, it is not a maneuver to be experimented with by the uninitiated. Improper technique can invite such startling surprises as an inverted spin or structural damage to the wings or control surfaces during a tail slide. Aerobatic training with a competent aerobatic instructor is absolutely essential before performing any aerobatic maneuver solo.

HORSE OF A DIFFERENT STALL The stalls that many pilots experience during their primary training may not be the type of stall they might encounter inadvertently with passengers occupying every available seat.

Pilots who learn to fly in one of the common, light, four-place airplanes will be doing most of their flying in the utility category weight condition. That is, there will be no passengers in the rear seats and no weight in the baggage compartment. Because of the way the airplane is loaded, all of their practice stalls will be performed at a forward c.g. A forward c.g. stall might be relatively mild because the elevators might not be effective enough to obtain a fully stalled angle of attack.

Because stalls are normally not practiced with passengers on board, pilots never experience the worst possible characteristics of a stall. Also, most airplanes are considerably more spin-resistant at a forward c.g. The wing will stall more completely at an aft c.g. because it is easier to obtain a greater angle of attack. See Figure 2.11.

Also, it is easier to effect a more complete stall in propeller-driven aircraft when power is being used. The increased airflow from the propeller as it flows over the tail results in increased elevator effectiveness, which enables one to attain a greater angle of attack. As a result, more of the wing becomes stalled, and a greater stall reaction is created. In addition, at low airspeed the propeller blast forces more air around the root section of the wing so that an effective boundary layer control and a delay in air separation are provided. This results in the wing tips stalling first, with a consequent loss of lateral control and a generally more dramatic reaction to the stall.

Wing-mounted engines and propellers, such as those found on multiengine aircraft, are most effective in increasing the airflow over a wing during slow-speed flight and in reducing the stall speed. Asymmetric power conditions on twin-engine aircraft not only affect yaw control, but roll control as well because of unequal lift. Obviously, the stall characteristics of a twin are greatly affected during single-engine operation. The wing with the dead

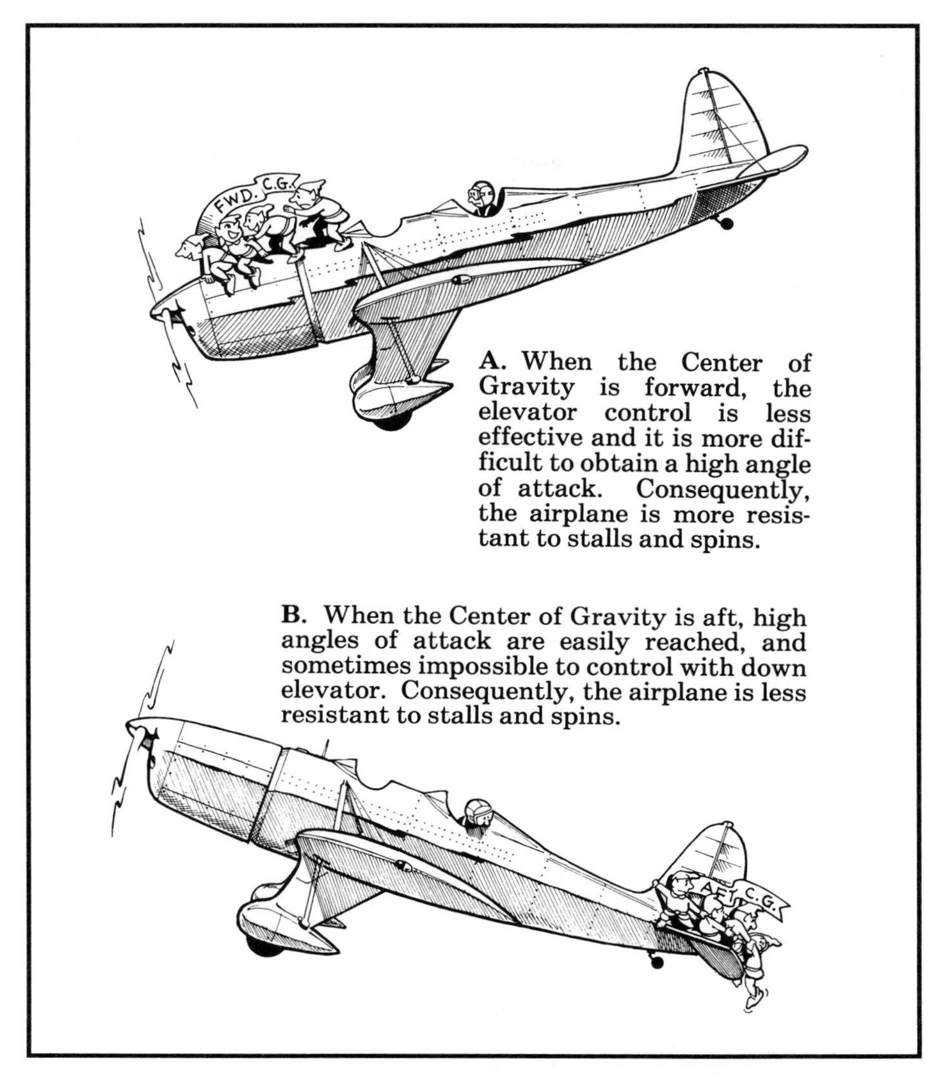

FIGURE 2.11 Effect of c.g. in obtaining stalling angle of attack.

engine will stall, while the wing with the good engine is providing good lift. The obvious result is an uncontrollable roll into the dead engine.

GROUNDED BY A STALL What are the stall characteristics of an airplane that is on the ground? If you didn't know that an airplane can be stalled with both main wheels still on the runway, you are lacking a very important bit of knowledge.

Except for the efficiency added to the wing due to its proximity to the ground, the aerodynamics of a wing are such that the wing performs the same on the ground as it does in the air. Lacking this understanding, many pilots have failed to become airborne when they expected to. If the wing is stalled before the pilot desires to lift from the runway, no amount of aerodynamic incantations will make the airplane levitate. I well remember one such experience.

The furnace of summer was turned high as I carried my flight gear to an awaiting F-80 jet fighter. With each step that I took in my trek to the airplane, the asphalt dipped sleazily, and I had to put forth greater effort.

I climbed aboard and attempted to buckle myself to the machine. Everything in the cockpit was hot. I quickly completed my cockpit check and wound the turbine with a flick of a switch.

I pushed the power lever well forward to roll the wheels out of the depressed asphalt and taxied rapidly to the runway as I closed the canopy and turned the air conditioner to full cold. I was anxious to get into the air and climb to cooler temperatures.

Aligned with the runway, I held the brakes and fully opened the throttle until the turbine developed maximum static thrust. I knew that the takeoff run would try my patience because I'd have to chase those skimpy air molecules longer and faster than usual to capture lift.

With a full load of fuel, my torch was heavy. I armed the tip tank release and released the brakes. The jet rolled on heavy tires. I watched the runway crawl slowly beneath the nose.

Slowly, my sun-baked steed gained speed and consumed the runway. I eyeballed the remaining distance and swept a glance past the airspeed indicator. Flight appeared within the realm of possibility.

I eased back on the stick to let the wings do their job. There was no response! In fact, I lost a little speed. I lowered the nosewheel to the runway and continued the roll, hoping for more airspeed. It was now too late to stop. I moved my thumb to the tip tank release button. Again, I eased the nosewheel from the runway, then lifted the nose a little higher. Beyond the fence ahead was a street loaded with automobile traffic. I pulled the nose a bit higher as I approached the fence. The wheels were still on the runway. My grip on the stick tightened as I squeezed the release button. As the plane rolled along the runway, the tanks gyrated wildly and then burst open. They spilled fuel onto the street ahead, but there was no fire. Nearly a ton lighter, the F-80 became airborne. I skimmed the fence and cars and slithered through a gap in a row of eucalyptus trees beyond.

Stalled while still on the runway? I doubt it, but I was close to it, mushing along on the backside of the power curve.

Flight instructors often fail to warn their students that the learning and practicing environment is often quite different from the real world of practical flying.

For example, taking off from a short, muddy field at a density altitude of 9000 ft is a whole hazardous world different from performing a "soft-field takeoff" from a 5000-ft, hard-surfaced runway at sea level.

Pilots should be trained in such a manner that they thoroughly understand the factors involved when they attempt standard training maneuvers under adverse conditions.

How skillfully an airplane is handled during a soft-field takeoff can determine whether or not the airplane will become airborne. There may not be enough power available to make the airplane do what the pilot desires under certain conditions.

During the initial phase of the takeoff roll, the drag caused by the wheels being mired in sand, mud, water, slush, or snow is considerable. The recommended technique is to get the nose wheel off the ground as soon as possible. This eliminates one-third of the drag caused by the runway surface. By continuing to hold the nose wheel off, the angle of attack of the wing will begin to provide lift that will reduce the surface drag being imposed on the main wheels. It is hoped that enough power will be available to overcome both the surface drag and the induced drag resulting from the high angle of attack of the wing. See Figure 2.12.

As speed is gained, the main wheels will tend to climb to the surface of the drag-producing mire. If the mire is slush or water, the wheels may begin to hydroplane. As this occurs, the angle of attack should be reduced to lessen the induced drag.

If, during the attempted takeoff roll, the rate of airspeed increase compares unfavorably with the distance remaining, the attempt should be abandoned.

Tall grass is a formidable drag producer. It may not be possible to raise the nose wheel high enough to clear it, and there is certainly no hope of the wheels being supported by its surface as speed is gained. The drag caused by grass increases dramatically with speed. I discovered this in a sensational way many years ago while looping a 36-hp Aeronca C-3 close to the ground. While looping over a field covered with tall grass, I recovered from the last half of the maneuver low enough to brush the wheels into the grass. As soon as the wheels touched, the Aeronca was snatched from the air! The propeller thrashed quarts of chlorophyll from its unusual surroundings as the windshield turned green. In brief seconds the small airplane had come to a complete halt. The airplane was undamaged, but I had to walk back to the airport to get my friends to help me move the machine to a field where the grass was mowed before a takeoff could be accomplished.

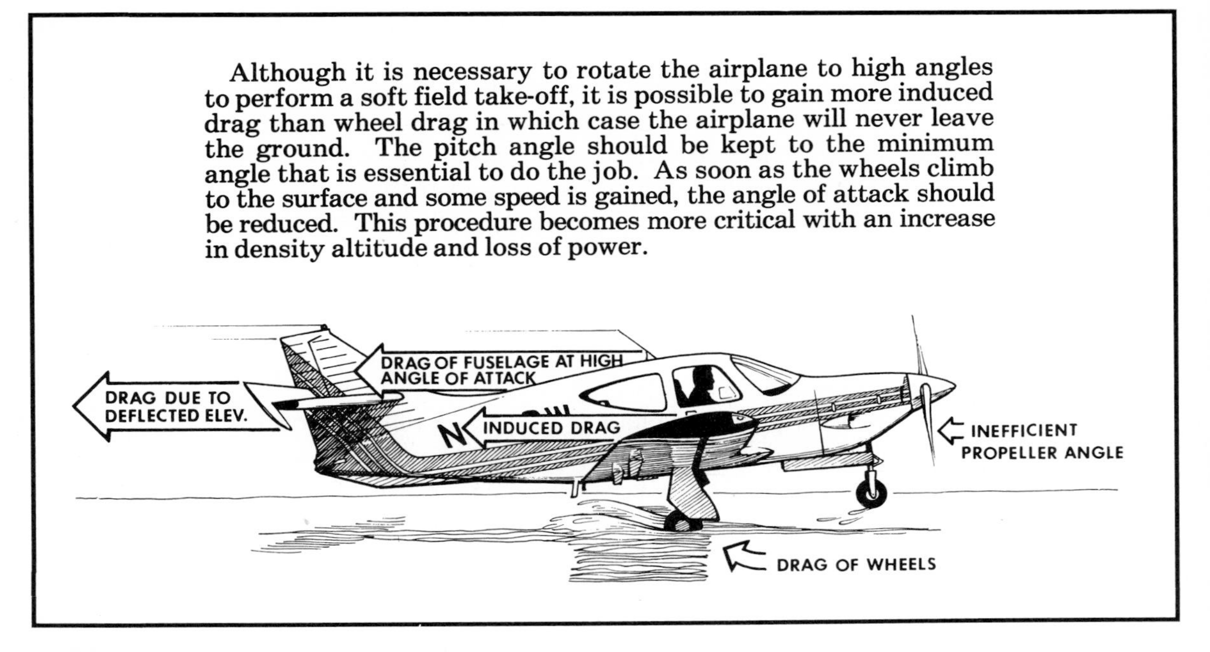

FIGURE 2.12 Soft-field takeoff.

MANEUVERING STALLS Wings are designed for lifting. How much lift they develop depends upon the area, airfoil, airspeed, air density, and angle of attack. In level flight the wings will be lifting the weight of the airplane plus the download being exerted against the horizontal stabilizer. If the airplane weighs 3000 lb, the wings will have to produce something in excess of 3000 lb of lift to carry it. As the airplane burns fuel, it becomes lighter, and the lift is reduced proportionally. The pilot will either reduce the angle of attack or the airspeed to maintain level flight.

However, as the airplane is maneuvered through turns, loops, dives, and rolls, the lift-to-weight ratio will vary considerably. During constant-altitude turns, the weight that the wings "feel" will vary with the angle of bank. This lift-weight ratio is expressed in *g*'s. In level flight the airplane is subject to the same gravitational pull that everything on earth is subjected to. This is referred to as 1 *g*, or 1 unit of gravity. However, during a 60° banked turn, the airplane will be subjected to 2 *g*'s. The wings will be lifting twice the weight they would be lifting in level flight. All curved flight (loops, turns, or combinations of both) creates its own gravitational effects. As the wings lift against the centrifugal force developed through these curves, the lift will vary according to the airspeed and the radius of the arc of flight.

Because the wings are feeling more weight during an increase in *g*'s, the stall speed increases.

A *g* is also referred to as a "load factor." In other words, if the airplane is being subjected to 2 *g*'s, it has a load factor of 2.

Figure 2.13 graphically illustrates the increase of *g*'s with angle of bank. Also, the increase of stall speed can readily be computed. This increase in

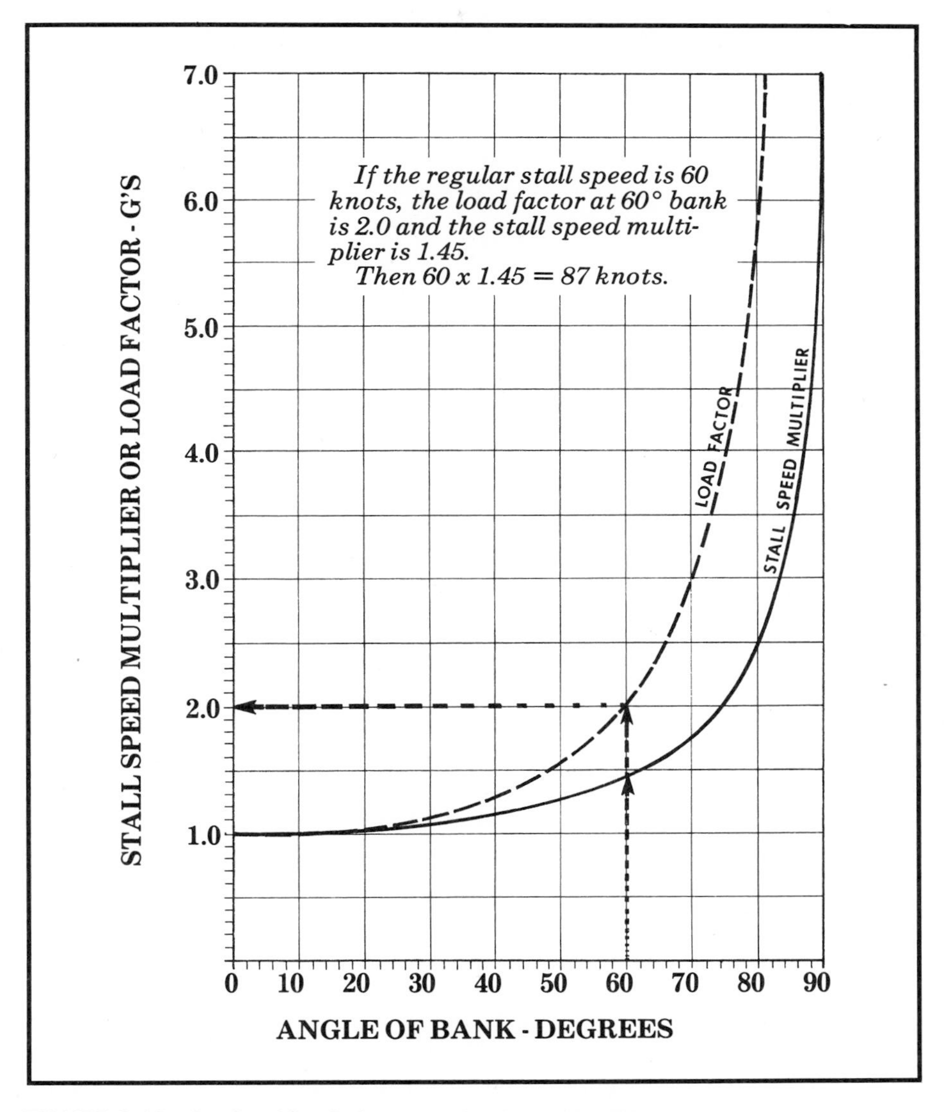

FIGURE 2.13 Angle of bank: Increase in *g*'s and stall speed.

stall speed is directly related to g's, and will stall at that speed, regardless of the attitude of the airplane.

LOAD FACTORS

Every airplane has a "limit load factor" that is expressed in both positive and negative g's. Airplanes operated in the normal category must have a positive limit load factor of at least 3.8 g's and a negative limit load factor of -1.52 g's. In the utility category, the requirements are 4.4 (positive) and -1.76 (negative). The aerobatic category requires 6.0 (positive) and -3.0 (negative).

The stall acts as a safety valve below a certain airspeed so that the airplane will stall before reaching the limit load factor. Within a specified maneuvering envelope, the pilot can maneuver abruptly without overstressing the airplane. When considering this maneuvering envelope, the pilot should also consider the age and condition of the airplane.

Obviously, an accelerometer (g meter) becomes a necessity if a pilot intends to maneuver an airplane near the limiting load factor, because deformation of the airframe becomes a very real possibility when the g values are exceeded.

While practicing for an air show at Van Nuys, California, I anticipated a square loop as I skimmed the runway in my Stearman. I squeezed the smoke trigger and sharply rotated the 450-hp biplane through the first 90° of the square. I climbed vertically, trailing a column of smoke behind the airplane. Then, at a greatly reduced speed, I quickly pulled the stick aft to square off the next corner. Flying inverted across the top half of the maneuver, I tossed my head back and looked at the runway. I seemed lower than usual. As I pulled sharply through the third segment, I realized I was too low to delay the fourth corner. I pitched to the highest angle of attack possible as the runway threatened to dismantle the airplane. Feeling stall buffet, I punched the stick slightly forward for a split second, then encouraged the Stearman through a lifesaving maneuver. I fully expected the wheels to hit the runway as I mushed rapidly to its surface. It didn't happen, and I was soon climbing to a safer environment, still gripping the smoke trigger. As I looked back, the smoke swept the runway, twirling vortices on each side. Relaxing my grip on the stick, I breathed easier, but determined never to attempt a square loop that low again.

The four corners of a square loop are good examples of the variable amounts of lift and g's obtained while arcing through the same radius at diferent speeds. See Figure 2.14.

The first 90° segment is performed at high speed, and to perform a sharp "corner," maximum g's are pulled. At this speed it is easy to overstress the

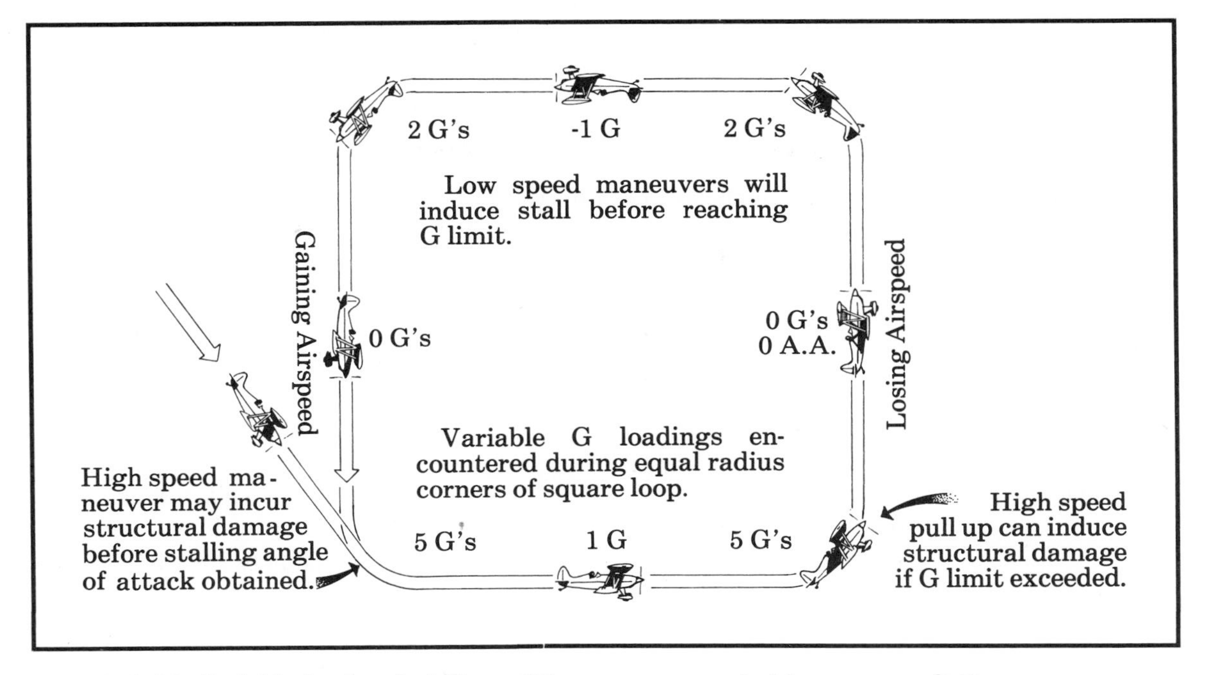

FIGURE 2.14 Variable load and stall conditions as represented by a square loop.

structure because it is possible to overshoot the *g* limitations before the wing can reach a stalling angle of attack.

The second 90° segment is accomplished at the lowest possible speed, but at the same sharp radius. The airplane would probably stall at less than 2 *g*'s at this corner because the reduced airspeed will not produce more *g*'s. The airplane is well within the structural maneuvering envelope at this point. The problem is to prevent a stall before the corner is completed. Also, after the second 90° segment is completed and the airplane is flown inverted to the next corner at low airspeed and at a high negative angle of attack, an inverted stall becomes a possibility.

The third 90° segment goes from a negative angle of attack, through 0°, to a high positive angle of attack. It is usually started well below the maneuvering speed (V_a), and although more *g*'s will be obtained than through the second segment, excessive *g*'s will not be possible. The stall will occur before obtaining limiting structural *g*'s.

The fourth segment may again become "*g* limited." In other words, so much speed may be gained during the vertical dive that if the pilot is not careful, he may overstress the airplane during the final 90°.

A stall can either be good or bad, depending upon where or how it occurs. It is good when it relieves the g-loads during high stress maneuvers, and bad when it occurs inadvertently at low altitude.

THE *V-n* DIAGRAM Figure 2.15 shows a V-n diagram; V stands for velocity, or airspeed, and the n for the number of g's. This is a fictitious diagram and doesn't represent any particular airplane, but it represents a common description of how the limiting factors are plotted so that pilots can quickly determine the operating envelope of a particular airplane.

Pilots should study the operating envelope and know the limitations of any airplane they plan to fly.

GROUND EFFECT—A BANEFUL BEGUILEMENT Years ago a very famous and exceptionally experienced pilot landed at Klamath Falls, Oregon, in an amphibian noted for its lack of performance. After picking up three friends, he taxied out for takeoff. The elevation of the airport was nearly 4100 ft. It was a very warm day, so the density altitude could easily have been around 7000 ft. He was concerned that the airplane might not get off the ground with four heavy men in it, so he decided that if the airplane didn't become airborne within a reasonable distance, he would abort the takeoff.

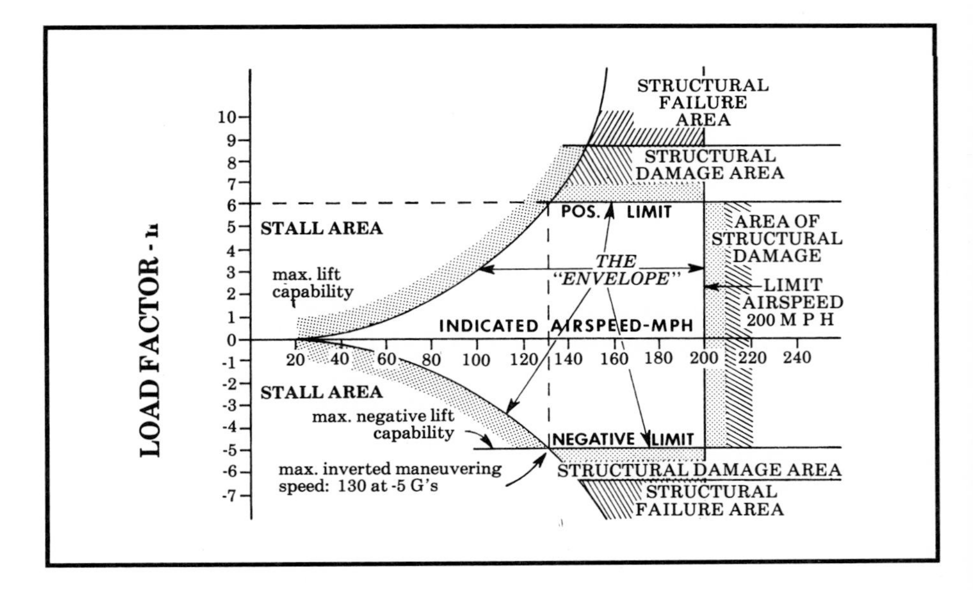

FIGURE 2.15 ***V-n* diagram.**

However, the airplane became airborne sooner than he expected, and he continued an attempted climb. The airplane climbed to about 30 ft and refused to climb higher. The pilot was now committed. There wasn't enough runway left to land on. Being a very good pilot, he lowered the nose and gave the airplane every chance to gain speed. He knew better than to point the airplane's nose above the power lines and trees ahead. I'm certain that he would have attempted to fly under the power lines if there hadn't been trees under them. He picked up as much speed as possible and at the last minute attempted to convert the speed into a zoom over the obstacles. It didn't work. One wing stalled and snagged the wires, and the airplane flipped over and dove into the ground. The crash killed one of the finest aviators of that era and two of the passengers. One man (a pilot) survived to tell what had happened.

The underlying cause of the accident was that the pilot was deceived by the performance of his aircraft while in proximity to the ground.

The ground-proximity phenomenon that is often responsible for bamboozling pilots into stall/spin accidents is the drag-reducing occurrence called "ground effect." Some pilots refer to it as a "ground cushion" because it produces an effect that feels like an extra amount of lift when the airplane is flown close to the runway. The term "ground cushion" suggests that the air is being compressed between the wing and the ground. This is not true. However, the ground does affect the airflow pattern ahead of and behind the wing, greatly reducing the preceding upflow and the receding downflow as the air approaches and leaves the wing. This, along with a considerable reduction in the size of the vortices trailing off the wingtips, results in a considerable reduction in drag. This drag reduction is measurable when the airplane is within a wingspan of the ground. If it has a span of 30 ft, the effect will develop when the wing is within 30 ft of the ground. The closer the wing is flown to the ground, the more noticeable the effect becomes. When a low-wing airplane skims the runway, its induced drag is reduced approximately 40 percent. See Figure 2.16.

It is now easy to understand why an airplane will perform more efficiently when it is flown close to the ground.

When the density altitude and the power available are only sufficient for the airplane to fly within ground effect, a dangerous situation exists. The ceiling of the airplane may be only as high as its wing span—perhaps not even that.

As an airplane climbs out of ground effect, two things occur that can be hazardous when its performance is seriously impaired. First, induced drag is increased. Second, because the downflow aft of the wing is at a greater down angle, it produces a larger angle of attack at the stabilizer, which results in

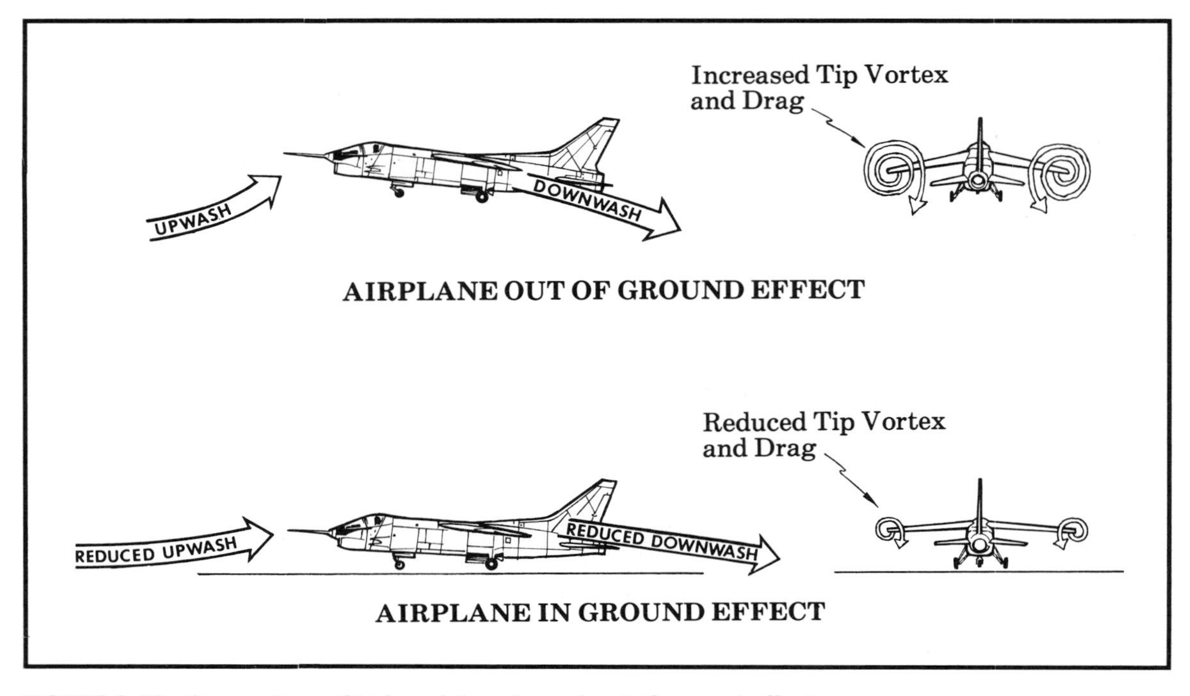

FIGURE 2.16 Comparison of induced drag, in and out of ground effect.

a tendency for the airplane to pitch up. Unwary pilots may find themselves in a stall or spin when this occurs.

If there is an obstacle ahead that is higher than the airplane can climb, an even more serious condition exists. If there is enough airspeed to turn away from the obstruction, a shallow turn should be attempted, keeping in mind that the turn itself creates more induced drag.

If there are power or telephone lines ahead, and it is impossible to fly over them, get as close to the ground as possible and fly under them—a far safer procedure under the circumstances.

It is far better to crash-land short of the obstacle while under control than to hit the obstruction near the top and dive into the ground.

STALLING EITHER SIDE OF THE WING An airplane can be stalled under either positive or negative *g* conditions. An inverted stall can occur with the airplane inverted or right side up. Conversely, an erect stall can occur while erect or inverted. It is possible to enter an erect or inverted stall from any attitude.

During positive g conditions, the airflow separates from the top surface of a wing when it stalls. During negative g's, the flow separates from the bottom surface. See Figures 2.17 and 2.18.

The airplane may exhibit different characteristics between erect and inverted stalls. For example, if the wing is twisted to wash out the tips in order to induce a stall on the root section first, rolling the wing inverted will

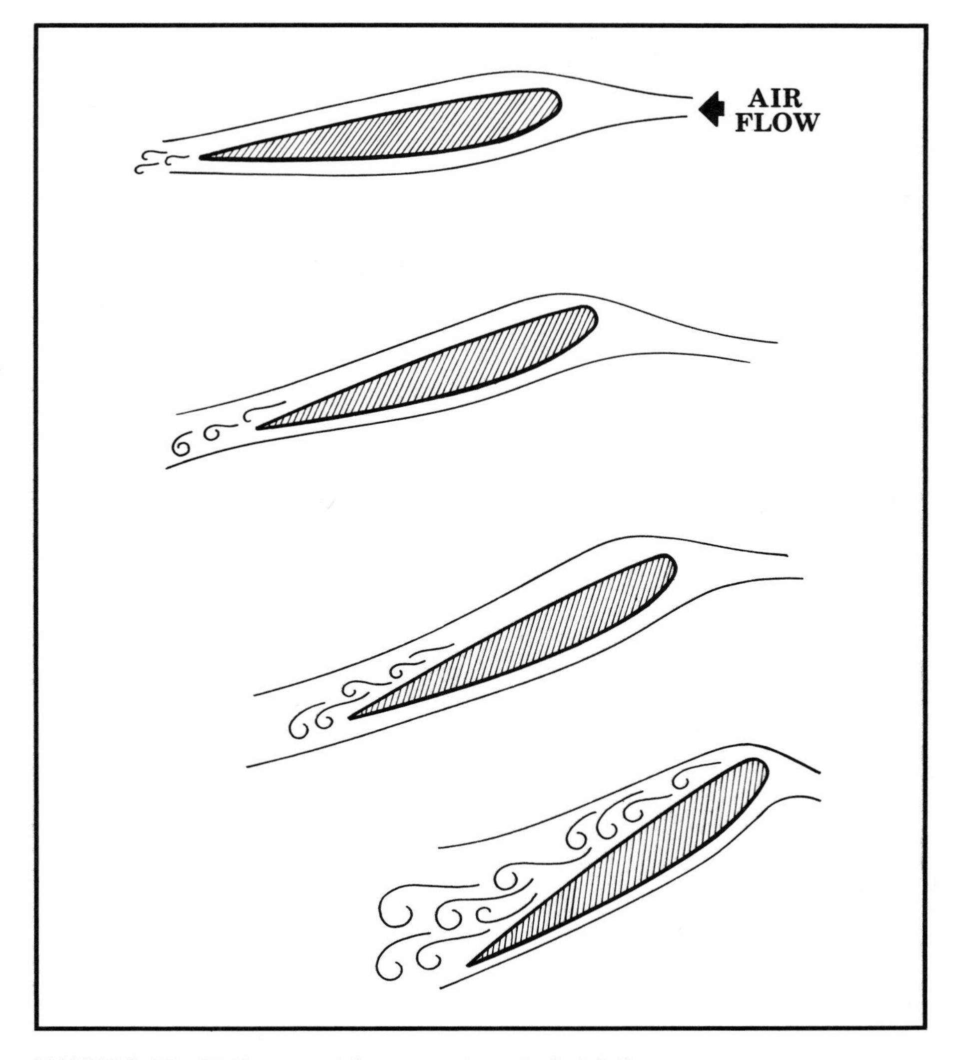

FIGURE 2.17 Stall progression on an inverted airfoil.

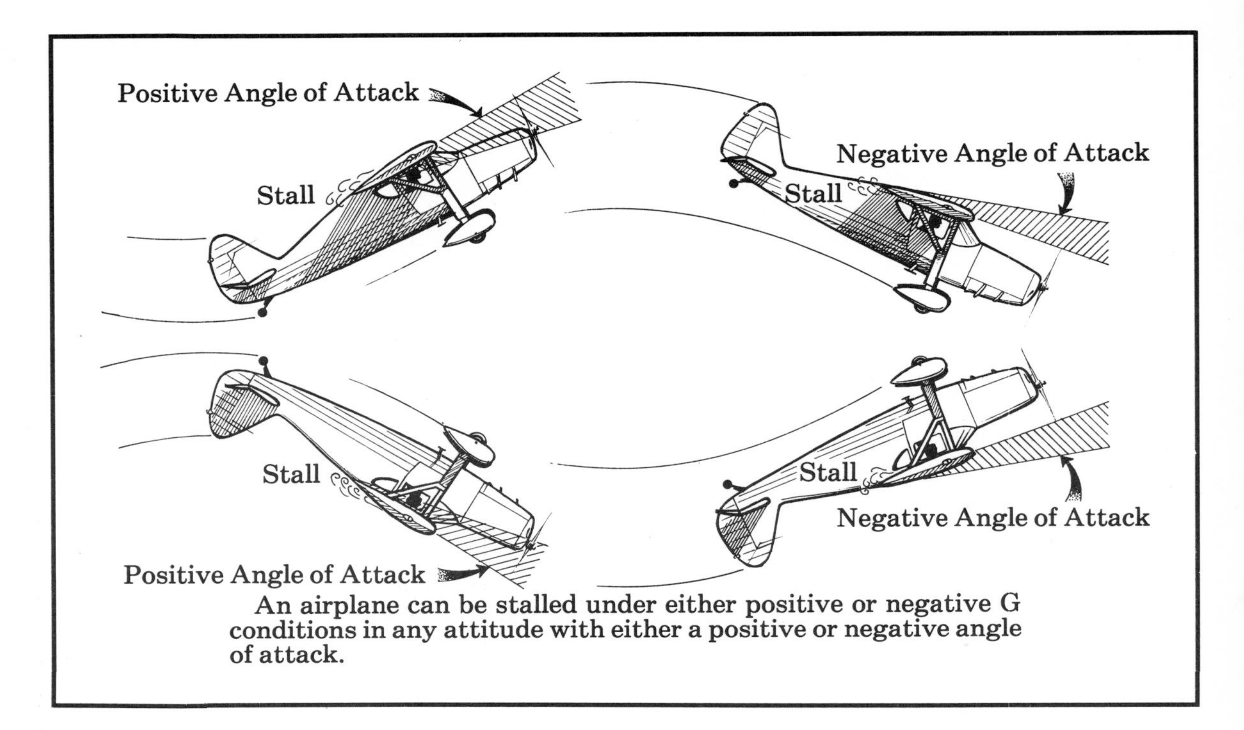

FIGURE 2.18 Positive and negative stall conditions.

produce the opposite effect. During a negative angle of attack, the wing tips will be washed in and, as a result, will stall before the root section. Also, unless the airfoil is symmetrical, the inverted camber will produce a change in stall characteristics. The nose-down pitch reaction that is common to erect stalls may not occur during inverted stalls. In fact, it is not uncommon for inverted stalls to pitch up during the initial phase of the stall.

VARIATION IN STALL BEHAVIOR The behavior of an airplane during a stall may vary in relation to how it is being maneuvered. For example, there can be a difference between a stall performed at 1 *g* with wings level and a stall encountered during a turn. The geometry among level, descending, and climbing turns is different enough so that the airplane will behave differently when stalled while performing these various types of turns.

During a level, coordinated turn, once the bank is established, the airplane will continue to turn about the yaw axis and pitch upward about the pitch axis. It will not be rolling about the roll axis. When a stall is encountered in a level turn, the reaction will normally be very little different than during a wings-level stall.

During a descending turn, or spiral, in addition to pitch and yaw, the airplane will be rolling about the roll axis in the direction of the turn. As the airplane rolls, it induces an upflow of air into the descending wing. This results in the descending wing having the greatest angle of attack. If a stall is encountered, the airplane will likely roll into the turn.

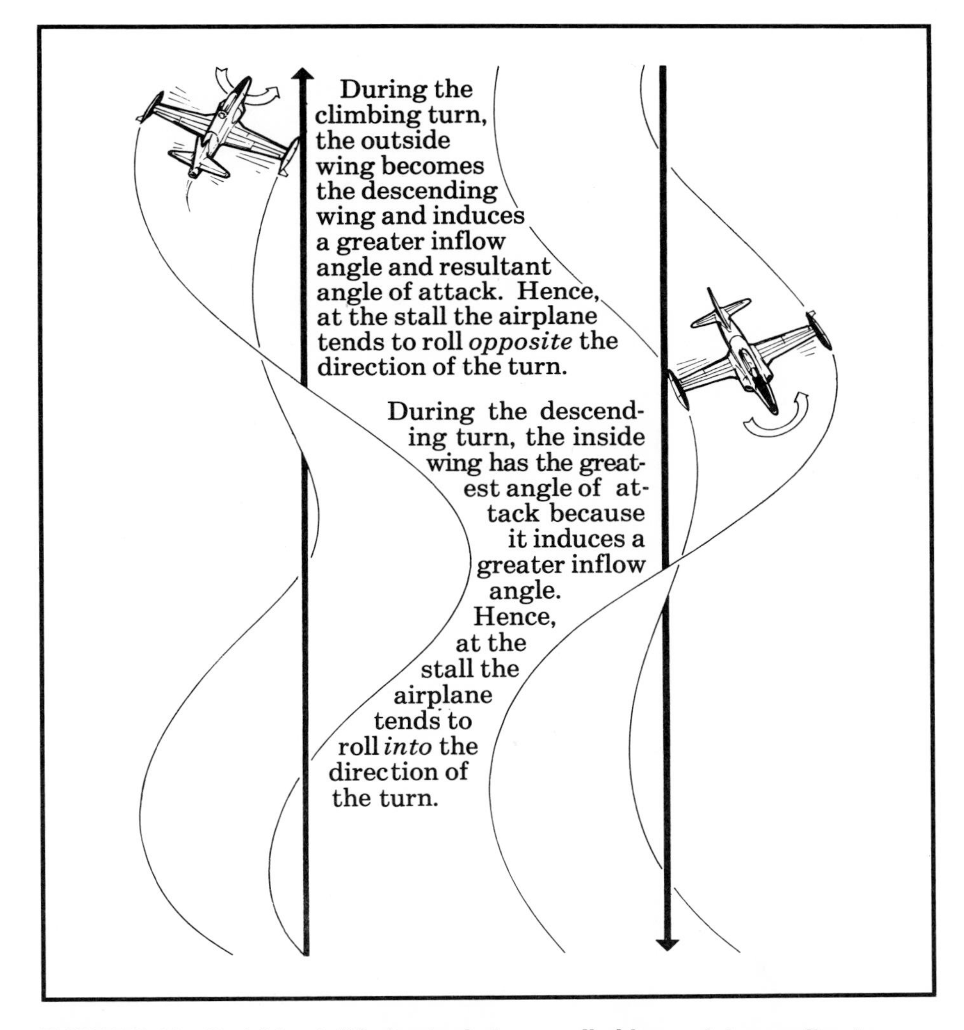

FIGURE 2.19 Variable stall behavior between climbing and descending turns.

During a climbing turn, the airplane is still moving about all three axes, but strangely enough it rolls *opposite* to the direction of the turn. This results in the high wing having the greatest angle of attack. When a stall occurs during a coordinated climbing turn, it will more than likely roll opposite to the turn, or "over the top."

In order to clearly visualize the roll behavior during climbing and descending spirals, obtain a hand-held toy airplane and go through the patterns of the level, descending, and climbing turns. Make the climbing and

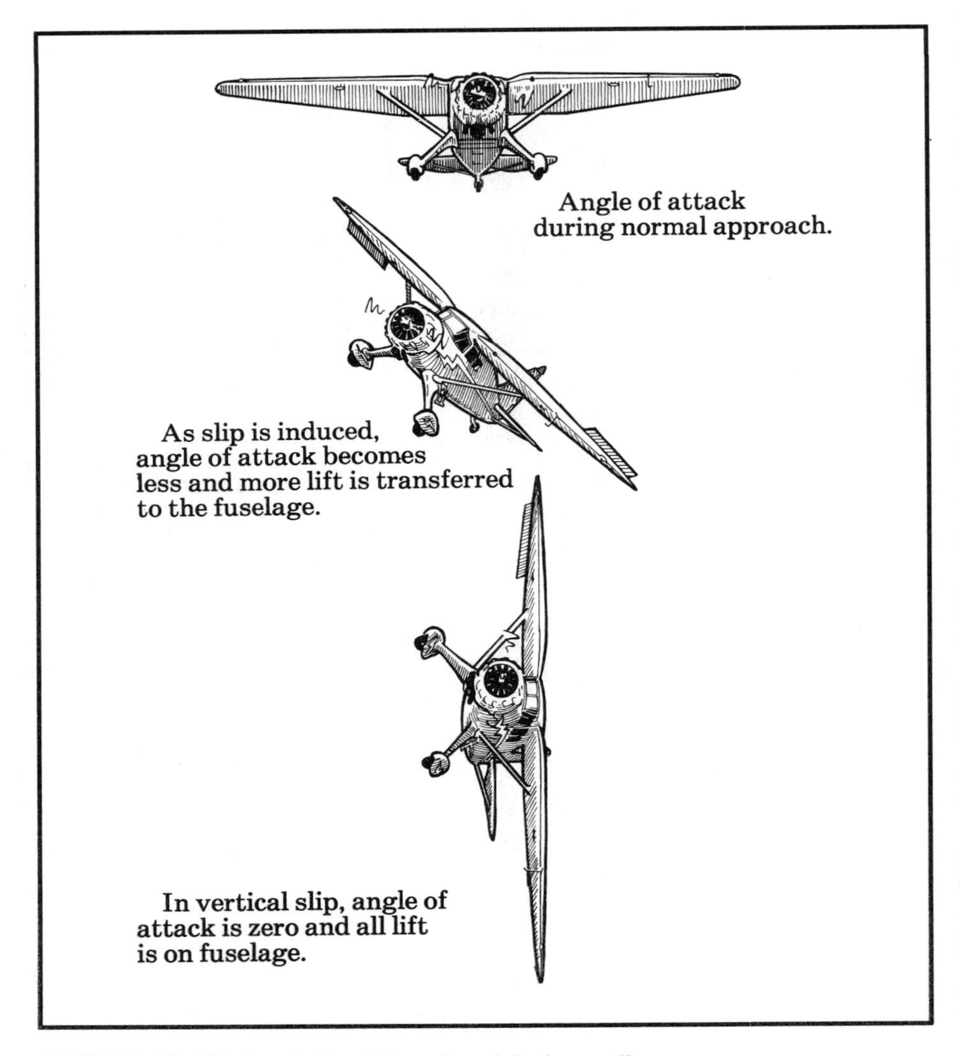

FIGURE 2.20 Angle of attack is reduced during a slip.

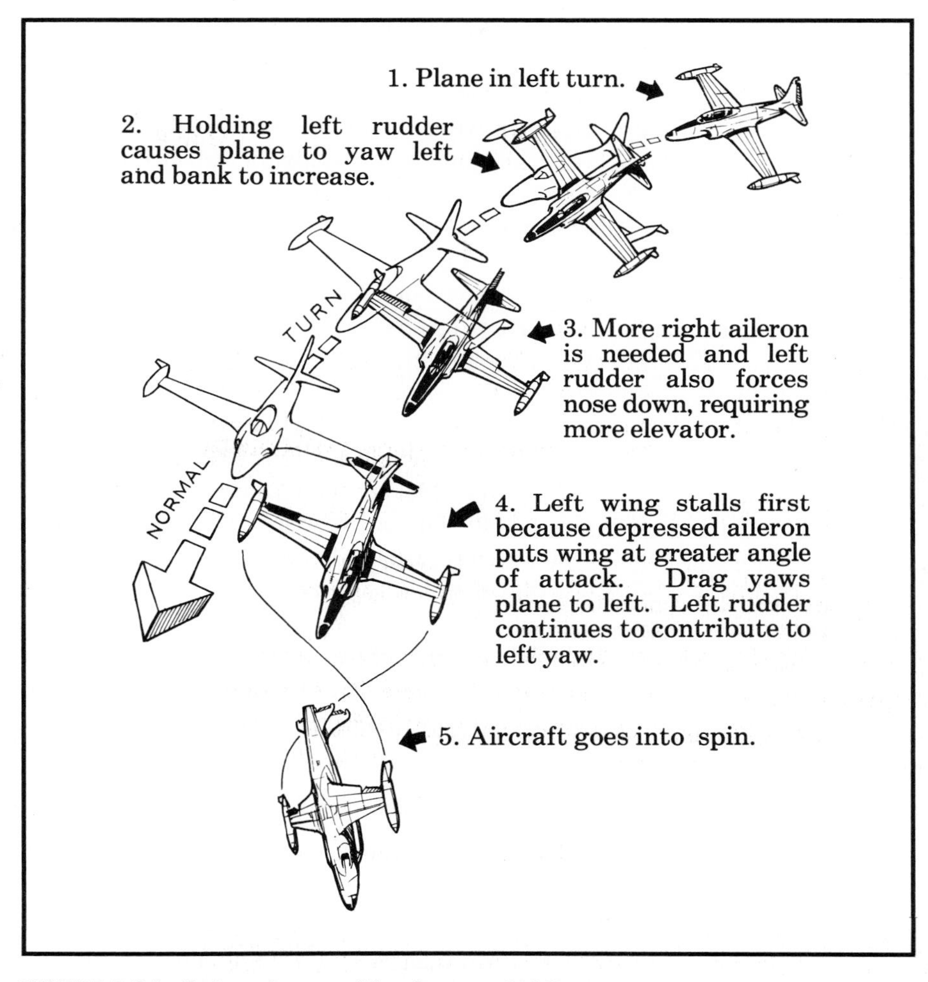

FIGURE 2.21 Spin entry resulting from a skidding turn.

descending turns exaggerated. You will notice that you must constantly roll the airplane into the descending spiral and away from the upward spiral. See Figure 2.19.

SLIPPING AWAY THE STALL One of the most difficult maneuvers from which to enter a stall or spin is during a slip, particularly a severe slip. Not that a stall cannot be entered if there is enough elevator authority; it most certainly can. But during a true wing-down slip, stalling is more difficult.

As the airplane is rolled into a slip and opposite rudder is applied to maintain a straight flight path, the lift is gradually transferred from the wing to the fuselage. Visualize an airplane being rolled until the wings are straight up and down while flying on a straight line. In this case there would be no angle of attack and no lift. Only the fuselage would be lifting. However, when the wing is not held down by use of the ailerons, a skid develops. Stalls and spins are entered quite easily from a skid. See Figures 2.20 and 2.21.

During a skid, the wing toward the rudder being used tends to drop, increasing its angle of attack. As the pilot applies opposite aileron to prevent the bank from steepening, the depressed aileron on the descending wing effectively further increases the angle of attack. During the bank, the excessive rudder also causes the nose to drop. The pilot pulls the stick/wheel aft to raise the nose. This further increases the angle of attack. Because of the skid, cross flow develops across the wing, blowing toward the low wing. The boundary layer thickens and causes premature separation to take place, inducing a stall. The airplane rolls in the direction of the rudder being held, or toward the low wing.

SUMMARY

1. Developing safe flying skills involves the development of deeply ingrained reflexes that respond correctly to every condition of flight.
2. The recognition of and early response to "mushing" are as important as responding promptly to stalls and spins.
3. Like the stall, mushing can occur at the higher speeds.
4. Stalls, spins, and mushes can occur in any attitude of flight. They should be practiced from unusual attitudes, especially from steep, nose-down positions.
5. A wing with inherently good stall characteristics can produce dangerous characteristics if a pilot displays poor stick and rudder work. The normal stall buffet may be eliminated and a tip-stall may occur. Easy spin entries result from this situation.
6. Any deformation of an airfoil, particularly on the top or the leading edge of the wing, caused by damage or the formation of ice, can deteriorate the stall characteristics.
7. A thin layer of frost can be deadly!
8. An airplane loses its resistance to stalls and spins with an aft c.g.
9. The use of power usually results in a deterioration of stall characteristics.
10. It is possible for a wing to be stalled or at such a high angle of attack while still on the runway that the induced drag will be so great that a

takeoff is greatly delayed or impossible. The effect of rolling drag caused by snow, mud, water, sand, or grass, combined with a high density altitude and a loss of power, can greatly add to the problem.

11. Stall speeds increase during turns or other conditions of maneuvering or curved flight. The stall can be a structural "relief valve" that prevents structural overload. However, if the airspeed is above the allowable maneuvering speed, structural damage will occur before the stall can take place.
12. Ground effect can deceive pilots. Its effect during takeoff can fool pilots. They may think the airplane will perform better than it is capable of once it climbs away from its effect.
13. An airplane can be stalled under either positive or negative *g* conditions. Inverted stall and spin characteristics can be different from those experienced while erect.
14. Turning stall characteristics can vary, depending upon whether they occur in level, climbing, or descending turns.

3

"The probability of survival is equal to the angle of arrival."

-P. T. Barnstorm

WHAT IN THE WHIRL IS A SPIN?

Although I learned to fly in an OX-5 Eaglerock biplane and put in most of my early flying time in a variety of OX-5–powered airplanes, I was afraid to spin them because I didn't know what to expect.

I watched while another student entered a spin in the Eaglerock and spun it all the way to the ground. However, it was such a flat spin, and the rate of descent was so slow, that he walked away from the wreck with only a bump on his head.

It wasn't until the 40-hp Taylor Cub came upon the scene that I worked up enough courage to try a spin. The Cub had an honest reputation concerning spins, and I decided it was an ideal airplane for me to experiment with.

My instruction in spins consisted of a talk with a pilot who had spun the Cub. He told me how to get into the spin and how to recover from it, then assured me that I would not experience any difficulties.

I'll never forget the experience. I climbed to 2000 ft (high for a Cub) in the vicinity of the old Telegraph and Atlantic Airport in East Los Angeles. I felt apprehensive as I throttled back and approached a stall; I was haunted by all the hangar stories I had heard about spins. I was uneasy, but I was programmed to react when the stall occurred.

Suddenly the nose pitched down and the wing dropped. I pulled the stick hard aft and pushed the rudder toward the low wing. Instantly the nose stirred the scenery below into a spinning panorama of twisted streets, railroad tracks, and distorted hangars. It was a shocking experience. I was glad that I was not experiencing this for the first time during an inadvertent spin in the traffic pattern! I was thankful that I was high. I pushed hard opposite rudder and shoved the stick forward. The Cub obeyed instantly and the spin stopped. I pulled from the dive into level flight.

Whew! I needed to think about this for a minute. As I climbed for more altitude I thought through what had taken place. Frankly, I was a bit shaken. I wished I had someone with me who was skilled in spins. As I thought about it, I realized that I had not experienced any physical loads such as I had experienced on a roller coaster. In fact, there was less of a sensation than that felt on a steep turn. Most of the reaction was from what I had seen and the unusual attitudes—the nose pitching nearly straight down, momentarily rolling inverted, and the ground spinning below me. I guessed that if I had closed my eyes, the sensation would have been minimal.

Climbing back to 2000 ft, I tried another spin. That one wasn't so bad, perhaps because I knew what to expect. The third one was a cinch. After several more, it was fun. It was just a matter of experience.

I certainly don't recommend the solo method of learning to recover from spins. Find a competent instructor; it will be well worth the money. You'll pump far less adrenalin, learn more, and most of all, you'll be a lot safer.

FAA DEFINITION

The FAA *Flight Training Handbook* describes a spin this way: " . . . an aggravated stall that results in autorotation. The airplane describes a corkscrew path in a downward direction. One wing is producing effective lift, and the airplane is forced downward by gravity, rolling and yawing in a spiral path."[1]

This is a good description of the pattern of a spin, but as we shall see later, the wing angles of attack that develop during a spin are well beyond the normal stalling angle that occurs at about 15°. Both wings are completely stalled during a spin.

SPIN OR SPIRAL?

Although the pattern of a spin resembles a spiral, they are two different maneuvers. However, there are times when they may be difficult to tell apart. The National Aeronautics and Space Administration (NASA) admits that it has difficulty separating the two terms.

A U.S. Navy publication says this: "The spin differs from a spiral dive in that the spin always involves flight at high angle of attack, while the spiral dive involves a spiral motion of the airplane at relatively low angle of attack."[2]

[1] *Flight Training Handbook,* FAA publication EA-AC 61–21, 1965, p. 115.

[2] H. H. Hurt, Jr., *Aerodynamics for Naval Aviators,* U.S. Navy Publication NAWEPS 00-80T-80, January 1965, p. 307.

When an airplane is in a fully developed spin, there is no doubt whether or not it is spinning. However, in some instances it is difficult to tell the difference.

In most cases, the pilot can tell if the airplane is spinning or spiraling by the airspeed. If the airspeed increases rapidly, the airplane is spiraling. If it remains near or slightly above the stall, it is spinning. The three stages of a spin are the incipient spin, the fully developed spin, and the recovery. See Figures 3.1 and 3.2*a* to 3.2*d*.

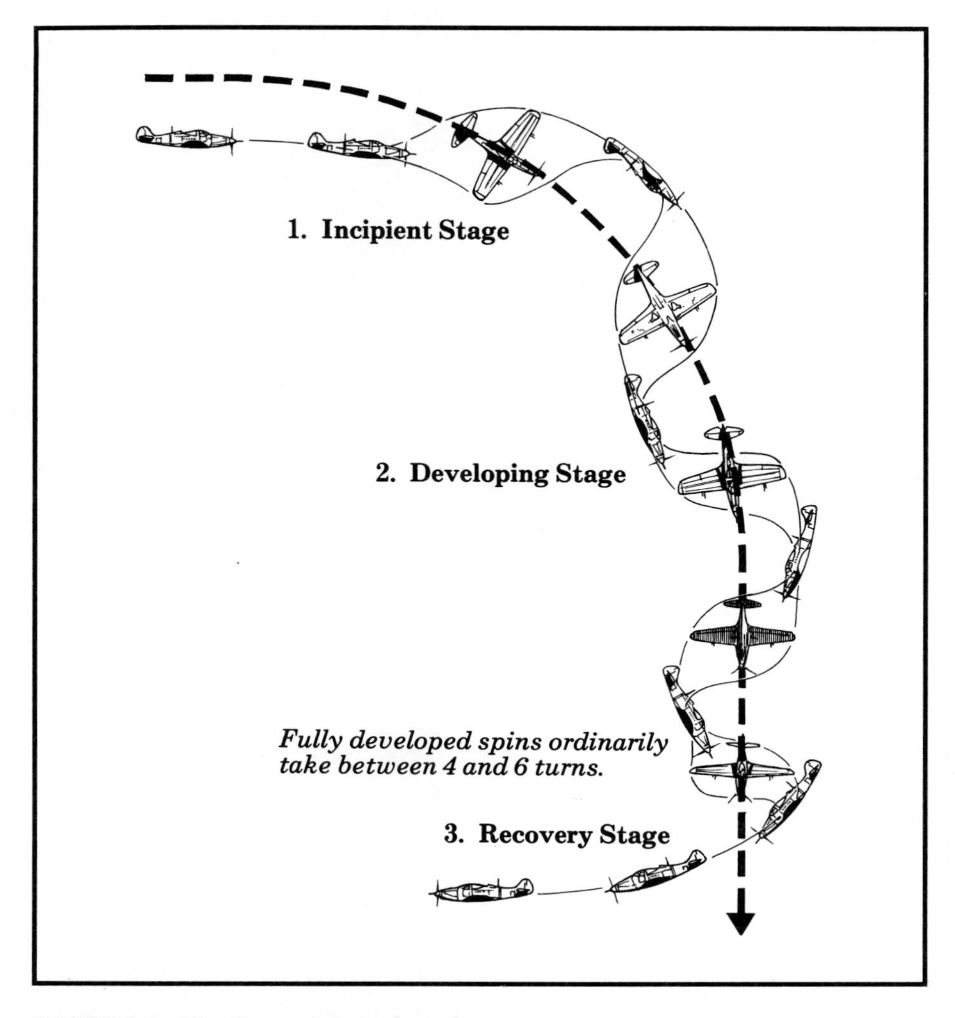

FIGURE 3.1 The three stages of a spin.

FIGURE 3.2*a* Spin entry from pilot's perspective.

FIGURE 3.2*b* Developed spin attitude from pilot's perspective.

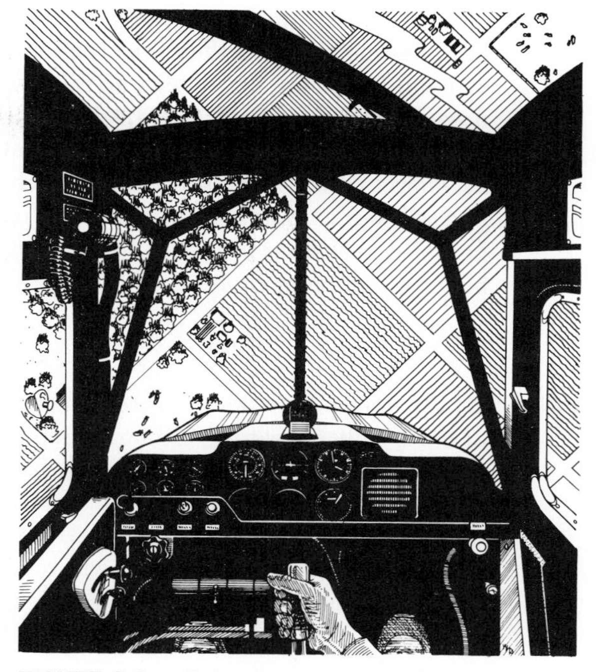

FIGURE 3.2*c* Spin recovery attitude from pilot's perspective.

FIGURE 3.2*d* Recovering from dive after spin recovery, from pilot's perspective.

INCIPIENT PHASE The incipient phase occurs during the asymmetric stall, or when one wing becomes more stalled than the other and the airplane starts to roll. Autorotational forces are developing at this time. Still in the incipient phase, the nose pitches downward, but the spin axis is not quite vertical.

As the nose drops further and the rotational speed increases, the spin axis becomes vertical. The axis of rotation will start somewhere ahead of the nose and will work toward the airplane as the spin develops. The flatter the spin becomes, the closer to the c.g. the axis will move. The axis may also be offset in the direction of rotation.

DEVELOPED PHASE By one and one-half to three turns, the spin will take on the appearance of being fully developed, although, depending upon the airplane, it may take five or six turns to develop its full dynamics. This can be identified by a constant nose attitude and rate of rotation. The nose attitude will be determined when a balance between aerodynamic and inertial forces is established. The aerodynamic forces will tend to lower the nose, and the inertial forces will raise the nose.

RECOVERY PHASE The recovery phase starts when the pattern of the spin starts to change after the application of recovery controls. Normally, the rotational speed will decrease and the nose attitude will steepen. There may be a momentary increase in rotational speed during this phase, but it will quickly diminish. This phase ends when all rotation has ceased.

THE SNAP ROLL Although the spin is described as a descending maneuver, it can, like a stall, be entered from any attitude. The snap roll is closely related to the spin. However, it is entered at a higher speed and is usually performed on a horizontal plane. Of course, if the rotation continues long enough, snap rolls, left to their own devices, will eventually enter a descending path.

Although the stall is the catalyst through which the spin is generated, the snap roll is a practical demonstration that the stall and the spin can take place simultaneously. Sometimes pilots have the impression that a spin can only occur after the airplane has penetrated aerodynamic stall buffet and they have had plenty of warning that a spin might develop. Very often, the conditions that result in inadvertent spins preclude buffet warning and the first indication of trouble is the spin itself.

SLOW-SPEED FLIGHT During spin practice you will discover that an airplane will enter a spin more easily if entry controls are applied just prior to the 1-*g* stall speed. The pilot who waits until the airplane is fully stalled before applying the rudder and up elevator necessary to induce spinning may find the controls relatively ineffective in producing an immediate spin.

The importance of proficiency in flying an airplane just above the normal 1-*g* stall speed is obvious: spins develop more readily from speeds slightly above the stall.

A DESCRIPTIVE RIDE Hop aboard my dual Whirling Whammy with me and let's explore the motivating mechanisms of a spin. We'll climb to a safe altitude (more will be said about this later), make a few clearing turns to make sure we will not conflict with other traffic, and have at it.

Although I want to make it clear from the outset that spins can be entered from any maneuver, from any attitude, and from relatively high speed, we're going to enter this spin from the approach of a normal stall (with which you should be familiar).

As I throttle back, I gradually pull the nose upward as the speed diminishes. I don't need to pull the nose up very high—just a little above landing attitude will do. When the airspeed reads about 5 miles per hour above the stall speed, I start to push on the left rudder (either rudder may be used). The reason I lead a little with the rudder is to permit the left wing to stall first. As the airplane is yawed to the left, the left wing descends (increasing its angle of attack) and it stalls slightly ahead of the right wing. As the airplane rolls to the left, I pull the stick/wheel fully aft. Both wings are now stalled as the nose pitches down. Interestingly enough, as the left wing drops, its angle of attack increases even more, resulting in a tremendous amount of drag. This drag also influences a strong pro-spin yaw and contributes to autorotation. See Figure 3.3.

AUTOROTATION It should be remembered that even beyond the stall, wings provide lift. More lift on one wing than the other will result in a rolling motion. However, it takes the drag of the descending wing to complete the autorotational forces. Although these forces are basic to the spinning of any airplane, other parts of the airplane can also add to the autorotational forces. For example, the forces acting upon the fuselage can also be pro-spin.

Now, with the rudder held full in and the stick held full aft, the aerodynamic forces that started the spin will cause continued spin rotation.

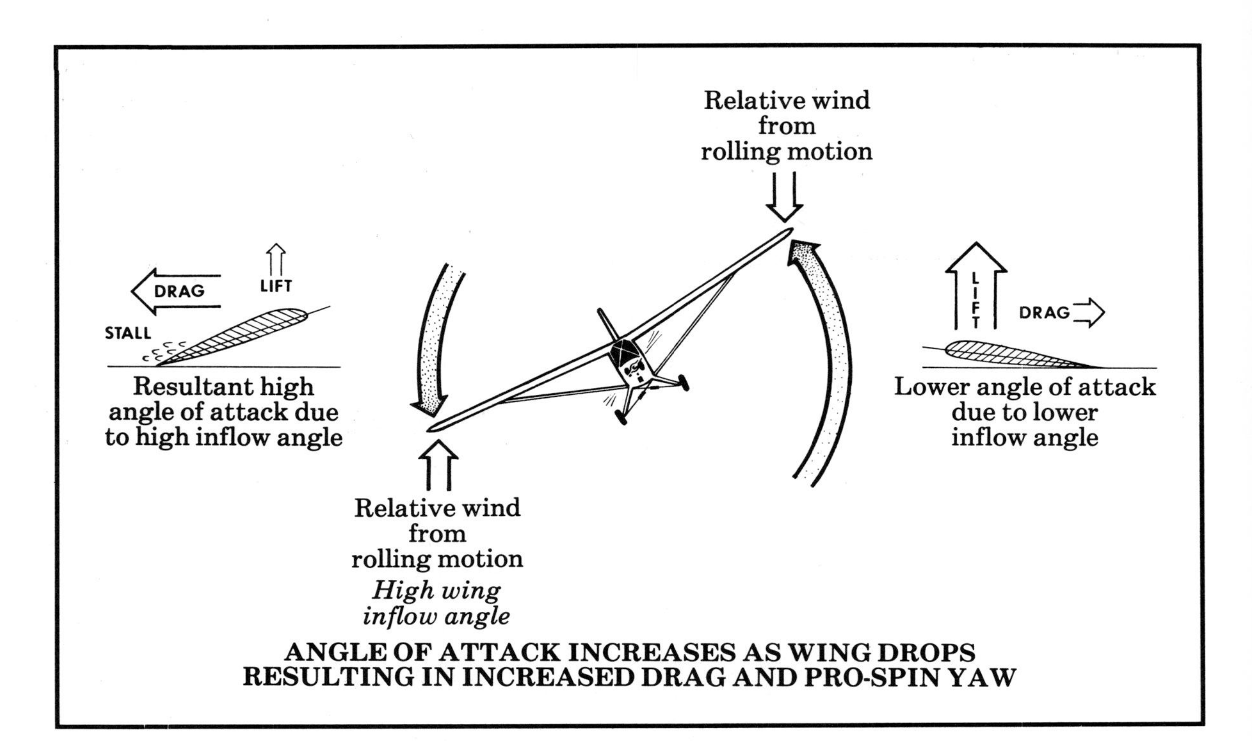

FIGURE 3.3 How autorotative forces develop during stall.

If you look at the airspeed, you'll observe that it hasn't increased much (depending upon the location of the Pitot and static sources) over the original spin entry speed, even though the nose is well down (the drag is terrific).

When the spin stabilizes, we will lose between 250 and 300 ft per turn. This is about average for most lightplanes. However, the first turn of rotation may involve the most altitude loss (when you consider that most inadvertent spins occur at very low altitude, this becomes very significant).

Observe that the ball is nearly centered, perhaps indicating a slight skid. This isn't always the case, but it indicates that we have a healthy spin involving both yaw about the vertical axis and roll about the longitudinal axis.

Notice that the *g* meter indicates only slightly more than 1 *g*. There is very little stress being imposed on the airplane.

THE RECOVERY Have you had enough? Okay, to recover I'm going to push hard rudder against the rotational direction; in this case, to the right. While applying rudder, I will hold the stick fully aft. For reasons that I will explain later, this provides the best aerodynamic effectiveness for the rudder and prevents other dynamic problems. After using the rudder, I briskly move the stick forward to about neutral (depending upon the character of the spin). We'll talk more about this later, but some airplane will require just relaxing the back pressure against the stick, and others will require full forward stick. The pilot's handbook for a particular airplane will reveal this information.

As the rotation stops, I neutralize the rudder. Now we are in a steep dive and the airspeed is increasing rapidly. I smoothly increase the back pressure against the stick and recover from the dive. I am careful not to again open the throttle until I am arcing upward toward level flight and the airspeed approaches the cruising range.

This is a spin. Nearly all airplanes are capable of exhibiting this phenomenon in various severities and patterns. It is an awesomely complex subject about which no one has all of the answers. However, enough is known about spins and the reasons that they develop to provide you with enough information to make you a more informed and, through practical application with a qualified instructor, a more skillful pilot.

FACTORS THAT RESIST SPINNING Like water, air resists any object that is thrust through it. Although air is the motivator of a spin, it also provides a resistance to rotation. The air will resist, to some extent, any tendency for the airplane to pitch, roll, or yaw. The amount of resistance will depend upon the density of the air and the size and shape of the area trying to be displaced through it. Thin air (high density altitude) does not offer as much resistance as denser air.

Like a weather vane, the vertical stabilizer tends to keep the airplane aligned with the relative wind. If the airplane is yawed right, for example, the stabilizer is subject to an angle of attack that creates a lifting force that swings the airplane back into the relative wind. See Figure 3.4. Also, if the nose of the airplane is pitched up or down, the horizontal stabilizer creates the longitudinal stability necessary to straighten the airplane with the relative wind. These stabilizing forces, particularly relative to directional stability, are important factors in resisting tendencies for the airplane to spin.

Although, as autorotation develops, the fuselage can contribute to autorotational forces, the vertical fin and fuselage can play a great role in resisting pro-spin forces. However, any part of the airplane, as it spins through the air, is subject to air resistance, which slows the rotation of the spin. The size of the airplane, its weight relative to its size, the density of the air it is spinning through, and the amount of aerodynamic forces at work to produce autorotation will determine how fast the airplane will spin.

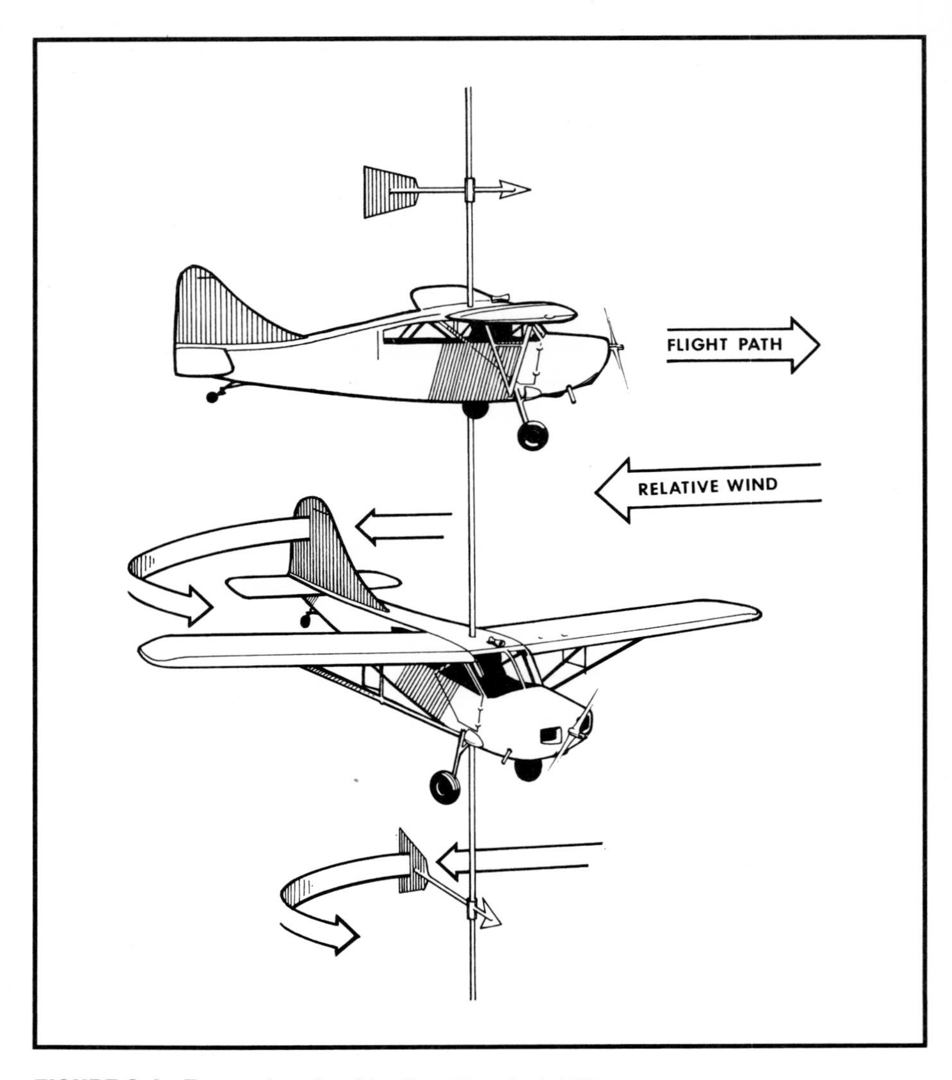

FIGURE 3.4 Forces involved in directional stability.

Strong directional stability contributes to spin resistance because it is the coupling of yaw in the direction of roll as the wing becomes stalled that excites a spin. An airplane with strong directional stability will tend to resist the yaw caused by the drag of the stalled wing.

Anything that limits the effectiveness of the elevator in pitching the wings to a stalling angle of attack results in making the airplane more spin resistant. Restricted up-elevator travel, for example, will make it more difficult to enter a stall and consequent spin. A forward center of gravity also greatly reduces the elevator effectiveness in driving the wing to high angles of attack. A forward c.g. makes an airplane more resistant to spin entries, but does not necessarily improve the spin and recovery characteristics. I have flown airplanes that spin worse at a forward c.g. than at an aft c.g. NASA's spin research pilot, Jim Patton, confirmed that he had experienced that same thing with NASA's modified Grumman Yankee.

High moments of inertia about any axis make the airplane more spin resistant. However, once a spin is entered, it may become more difficult to stop. An airplane that has considerable mass along the wings, in the form of fuel, engines, or structure, may resist the initial roll-off tendency, but once the spin begins to develop, it becomes more difficult to stop. When a high amount of rolling inertia is subject to the upward pitching moment of a spin, the effect of gyroscopic precession converts the rolling moment to yaw and the spin flattens.

Many modern aircraft have a certain resistance to spins. However, nearly all of them will spin if the conditions are right. These situations often occur inadvertently through circumstances that seem to incapacitate the ability of some pilots to control the airplane properly.

Although airplanes may have a resistance to spinning, the pro-spin forces can be greater than the anti-spin forces. This will be our next subject.

PRO-SPIN FORCES The forces that promote autorotation must exceed the spin-resistant forces before a spin is possible. Spins are incited by aerodynamic reasons, and their breeding ground is the stall. Of all the culprits responsible for the initial exciter of autorotation, the yaw (skid or slip) is the worst. If a pilot would always prevent a yaw from developing during the stall, a spin would be impossible. Training in yaw prevention during the stall is of utmost importance.

CONDITIONING CALISTHENICS One of the practice maneuvers I use extensively is to have students fly an airplane into a power stall while they hold the stick/wheel full aft. They are then encouraged to prevent the nose from turning by using the rudder while holding the ailerons centered. As long as they are able to prevent the nose from turning, the airplane will not spin. It may roll violently one way or another and threaten to spin, but as long as they prevent a coupling of yaw with the roll, the airplane can do no more than threaten to spin. Training in these types of exercises would go a long way toward preventing inadvertent spins. Of course, some airplanes lack the rudder authority to prevent a coupling of yaw with roll. The use of ailerons to correct for a roll-off due to stall can create so much drag on the stalled wing that it yaws the airplane into a spin in spite of full rudder being held against it. As previously explained, the descending, stalled wing develops such a high angle of attack and consequent drag that it becomes the predominant autorotative force. I have flown airplanes that would spin against the opposite rudder as long as an aileron was being used against the direction of rotation. The use of opposite aileron can greatly increase the drag on the inside wing (the lowering wing) of a spin.

Any condition that produces a strong yawing effect during the stall will excite a spin. For example, if left yaw is not corrected during a climb (due to P factor, torque, or asymmetric propeller loading), power alone is enough of a factor to provide sufficient yaw at the stall to incite a spin. If the pilot uses the ailerons to maintain a wings-level condition at this point, the drag of the low aileron will further yaw the airplane to the left. I frequently demonstrate this maneuver to students and show that the airplane will enter a spin with my feet off the rudder pedals.

The pro-spin aerodynamic forces are as follows:

1. An asymmetric stall (one wing more stalled that the other) produces a roll-off in one direction or another. The descending wing induces a greater air inflow angle, which results in a high angle of attack on that wing. The drag resulting from this extremely high angle of attack is tremendous and couples yaw into the roll.
2. Holding the rudder in the direction of the spin and the stick fully aft assures a combined stall and yaw. This results in continued rotation.
3. Fuselage forces produced as the rotation continues (especially if the spin flattens) can contribute to autorotation.
4. The use of aileron opposite to rotation can produce enough drag to create a pro-spin force.

Airplanes with good spin-recovery characteristics will eventually stop spinning as the rudder and elevator forces are released and the power is reduced to idle. The anti-spin forces will predominate and the spin will ultimately cease. However, this may take several turns and considerable altitude. Some airplanes will stop spinning almost instantly, and some will respond slowly and stop spinning after several turns.

AILERONS

You will want to learn to stop the spin as quickly and as effectively as possible. This requires brisk use of rudder opposite to rotation, followed by moving the stick/wheel forward to the position recommended by the manufacturer. The ailerons are usually kept centered during the recovery, but on certain airplanes and under certain conditions, the use of ailerons during a recovery, either toward or against the spin, may be recommended by the manufacturer.

The brisk use of opposite rudder is a recommended procedure on all general aviation airplanes. However, the use of elevators or ailerons may vary from airplane to airplane. I have found that the way the airplane responds to the rudder will give a clue as to how much forward stick will be needed. If rotation stops immediately upon using the rudder, I know that all that is needed is a release of back pressure against the stick. If the airplane responds very slowly to the rudder, it usually means that I will briskly move the stick to neutral. If it does not respond at all to the rudder, I know that I may need full forward stick.

More about Ailerons The manufacturer of one light civil training airplane recommends that opposite aileron be held to promote spinning. The airplane is so spin-resistant that it is the only way it can be spun. By using opposite aileron, a greater effective angle of attack is induced on the downgoing wing. The increased drag assists the rudder in holding the airplane into the spin.

However, on airplanes that enter spins more readily and easily continue rotating, the use of opposite aileron can flatten the spin, increase rotational speed, and make recovery difficult, and in some cases impossible.

Ordinarily, the use of aileron with, or in the direction of, the spin will speed up the rotation but will not flatten the spin. In most cases the spin angle will be steeper. The increased rotational speed may result in a delayed recovery.

Usually, the use of ailerons is not recommended during spin recoveries. The use of opposite aileron to stop rotation may increase the rotational speed and delay the recovery. However, the use of aileron in the direction of rotation might be used to advantage during recoveries in airplanes with predominantly heavy fuselages, or during recoveries from flat spins.

POWER—THE PROPELLING PERPLEXER The use of more power in propeller-driven aircraft increases the airflow over the rudder and elevators, thereby inducing a more positive spin entry. However, as soon as autorotation starts, the power should be retarded to idle. A slight amount of power may be necessary in some cases to keep the engine running. Engines with carburetors frequently experience difficulty because of the centrifugal loads disturbing the fuel level in the float chamber. However, fuel-injected engines give less trouble in this regard. Keep in mind, however, that too much power may aggravate the spin characteristics. It will increase the rotational speed, flatten the spin, and make the recovery difficult.

PROPELLER PRECESSION The whirling mass of a propeller can influence spin characteristics through gyroscopic and precessional forces developed when the airplane is displaced about the yaw and pitch axes. The size and weight of the propeller, in relation to the size of the airplane and the speed with which the propeller is turning, will determine the influence of the propeller as a gyroscope.

As with other factors that influence spins, the gyroscopic involvement is complex. The effect it will have upon the spin will depend upon the angle of attack of the spin, the amount of roll that is coupled with yaw, the rotational speed, and factors that tend to cancel the precessional effect.

Because the gyroscopic precessional effect of its clockwise-turning propeller (as viewed from the cockpit), the American airplane will display the following reactions (see Figure 3.5):

1. When the nose is pitched downward about the pitch axis, the nose will tend to swing to the left (tail-dragger pilots know better than to raise the tail too rapidly on powerful airplanes during the takeoff roll).
2. When the nose is pitched upward about the pitch axis at speeds above the normal left yawing tendency of the airplane during high power, the nose will yaw to the right about the yaw axis.
3. If the nose is yawed right about the yaw axis, the nose will pitch down.
4. If the nose is yawed left about the yaw axis, the nose will pitch up.

The amount of precessional reaction will depend upon the rate of movement about the pitch and yaw axes. The faster the rate, the greater the reaction.

The use of power as it influences propeller precession during spins will tend to produce the following effects:

1. Right spins (or snap rolls) tend to rotate faster. This is because an upward pitch induces a strong right yaw, which also influences the roll rate.
2. Left spins tend to flatten easier. This is because the left yawing motion creates propeller precessional forces that raise the nose.

The propeller precessional effect reacts in the same manner on inverted spins and snap rolls. However, because the direction of the maneuver

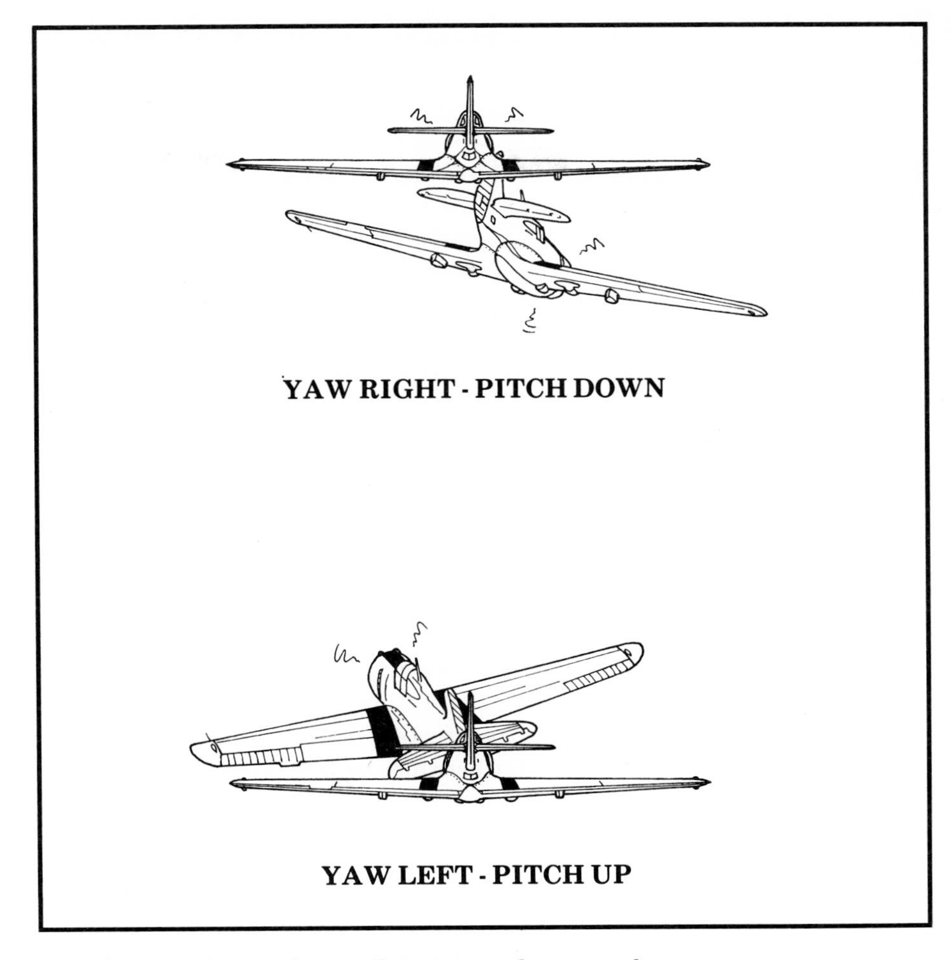

FIGURE 3.5 Effects of propeller gyroscopic precession.

appears different to the pilot than to someone watching it from the ground, the precessional reactions will appear different to the pilot.

A left inverted spin, as viewed by the pilot, is really a right spin as far as true rotation is concerned. So, just as right erect snap rolls, as viewed by the pilot, are easier to enter, the left inverted snap rolls and spins will rotate faster.

The above-mentioned considerations are only generally true. Some airplanes may have design considerations that outweigh the precessional effect of the propeller. Spin dynamics is an extremely complex subject, and the precessional effects may be very prominent in some airplanes and barely detectable in others. Airplanes that swing large, heavy propellers at high rpm will induce more precessional effect than airplanes that use small, light propellers at low rpm.

DESIGNS AND DYNAMICS AFFECTING THE DESCENDING DERVISH

There is no simple formula or rule of thumb that will predict how an airplane will spin and recover. The overall shape of the airplane, its density relative to the air that surrounds it, its mass distribution, and the location of the center of gravity will collectively determine its spin characteristics. In the final analysis it is the test pilot with steely eyes and sweaty palms who will determine the validity of the theory and research that apply to a particular design. However, there are some known factors that strongly influence spin characteristics.

Relative Density

Let's imagine that we have two feathers of the same aerodynamic shape and size. They have a slight twist to them that will result in a spinning action when they are released to fall to the ground. One of them is a real feather. The other one is formed from lead.

Now let's imagine that we're standing within a large cylinder that is totally void of air. We hold the two feathers at arm's length and let them fall. With no air resistance, they both fall at the same rate and land at the same time. Neither spins or has any aerodynamic reaction.

Now we step out of the cylinder and into the atmosphere. Again, we drop our two feathers. The lead one drops quickly to the ground without any evidence of being affected by the air. However, the real feather descends very slowly and immediately starts to spin. It quickly reaches a constant rotational speed and descends at a steady rate.

Now we take our two feathers several thousand feet high and drop them from a balloon. We observe them as they descend. The real feather acts just as before; it quickly reaches its constant rotational speed and settles at its slow, constant rate. The lead feather falls rapidly past the other feather and starts to rotate very slowly. Its rate of rotation gradually increases as it descends, until, finally, it is spinning much faster than the real feather. It is obvious that it would be more difficult to stop the spinning of the lead feather by aerodynamic means.

From the imaginary demonstration we can easily understand one of the factors that is of great importance to airplane spin characteristics: the density of the airplane relative to the density of the air. At higher-density altitudes, both feathers would spin faster. If we were to spin two airplanes of the same shape and size, but of different overall weights, the heavier airplane (the one with the greater mass) would accelerate more slowly, but would eventually spin faster and would be more difficult to stop.

Airplane Gyroscopic and Precessional Effects A spinning gyroscope tends to remain rigidly in its plane of rotation. However, if enough force is applied to any particular point in an attempt to tilt the gyroscope, it will react to that force 90° "downstream" from where the force is applied. This is commonly referred to as precession. See Figure 3.6.

As the mass of an airplane is spun through the air, the airplane behaves like a gyroscope. The combined spinning masses of the fuselage and wings, as they rotate about their respective axes, interact with one another to produce either pro-spin or anti-spin forces. How the mass is distributed will determine this effect.

An airplane will resist acceleration about any axis that has high inertia. In other words, if the fuselage is heavy, rotation about the pitch and yaw axes

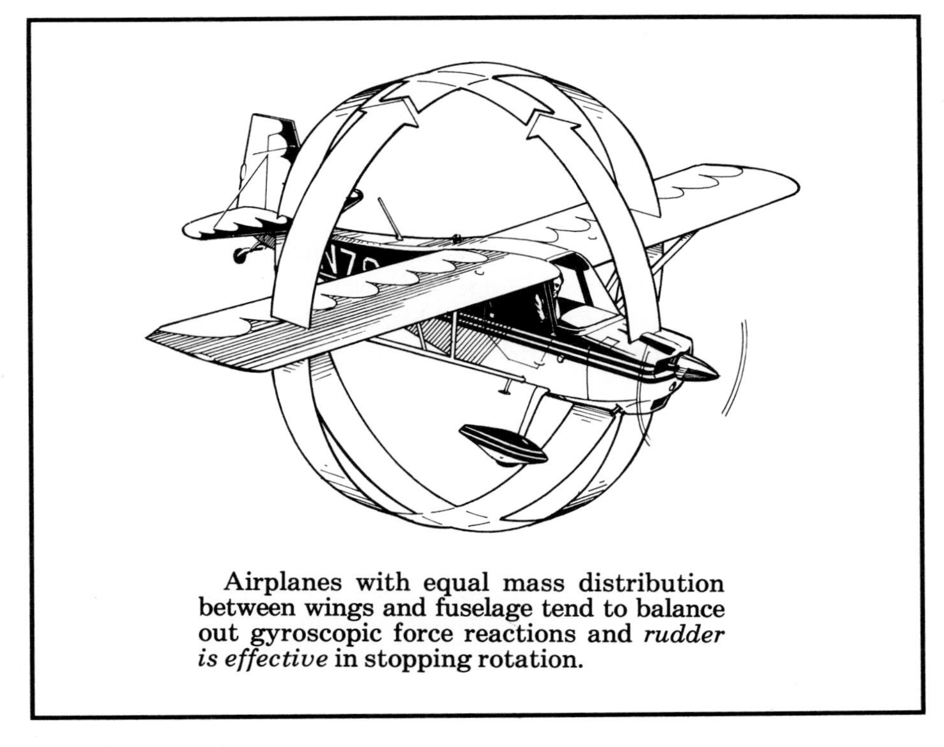

Airplanes with equal mass distribution between wings and fuselage tend to balance out gyroscopic force reactions and *rudder is effective* in stopping rotation.

FIGURE 3.6 Gyroscopic effects due to mass distribution along wings and fuselage.

will be resisted. If the wings are heavy, acceleration about the roll and yaw axes will be resisted.

A wing-heavy airplane, or an airplane that has considerable wing mass in the form of structure, fuel, engines, etc., may develop considerable rolling energy about the roll axis once a spin develops. If the nose is pitched upward about the pitch axis, the precessional effect of the rolling mass will result in a pro-spin yaw.

These gyroscopic effects can be used to advantage during recoveries. How the mass is distributed will frequently determine which control becomes most effective in stopping a spin. The first recovery control that should be applied is rudder. However, it may not be adequate in itself to provide a recovery. For the above-mentioned wing-heavy condition, down elevator would be the primary recovery control. The use of down elevator would provide a gyroscopic precessional effect that would become anti-spin as the nose is pitched downward.

Airplanes with an equal mass distribution along the fuselage and wings are said to have a "zero loading" and respond best to the rudder during a spin recovery. The average general aviation lightplane is a good example of an airplane with this mass distribution.

For a fuselage-heavy loading, the aileron is the primary recovery control. Fighter aircraft with short wings and long, heavy fuselages respond best to this control. However, these airplanes present their own set of spin problems. They develop combined aerodynamic and inertial problems that are interesting.

Imagine a tall, thin top spinning at high speed. Now imagine that you displace the uppermost part of the top slightly with your finger. At first there will be a slight wobble, and then it will wobble at higher amplitudes until, finally, it tumbles out of control.

Long fuselages not only produce inertial characteristics just mentioned, but also create aerodynamic reactions that are unfavorable for spinning. With so much aerodynamic area ahead of the c.g., the airplane can become directionally and longitudinally unstable at high angles of attack. An airplane with these characteristics is almost certain to display poor spin and recovery traits. The F-5, for example, has a flat, nonrecoverable spin, and it is not the only fighter airplane with spin problems. See Figure 3.7.

Once I was involved in some test work on a specially modified jet fighter. It had a very long nose and a large aerodynamic area ahead of the c.g. I was flying the airplane at low speed, investigating its stability prior to conducting stall tests. As I increased the angle of attack, I soon observed a reversal of elevator stick forces and a considerable diminishing of directional stability. I knew that if I proceeded into the stall, I would lose control of the airplane.

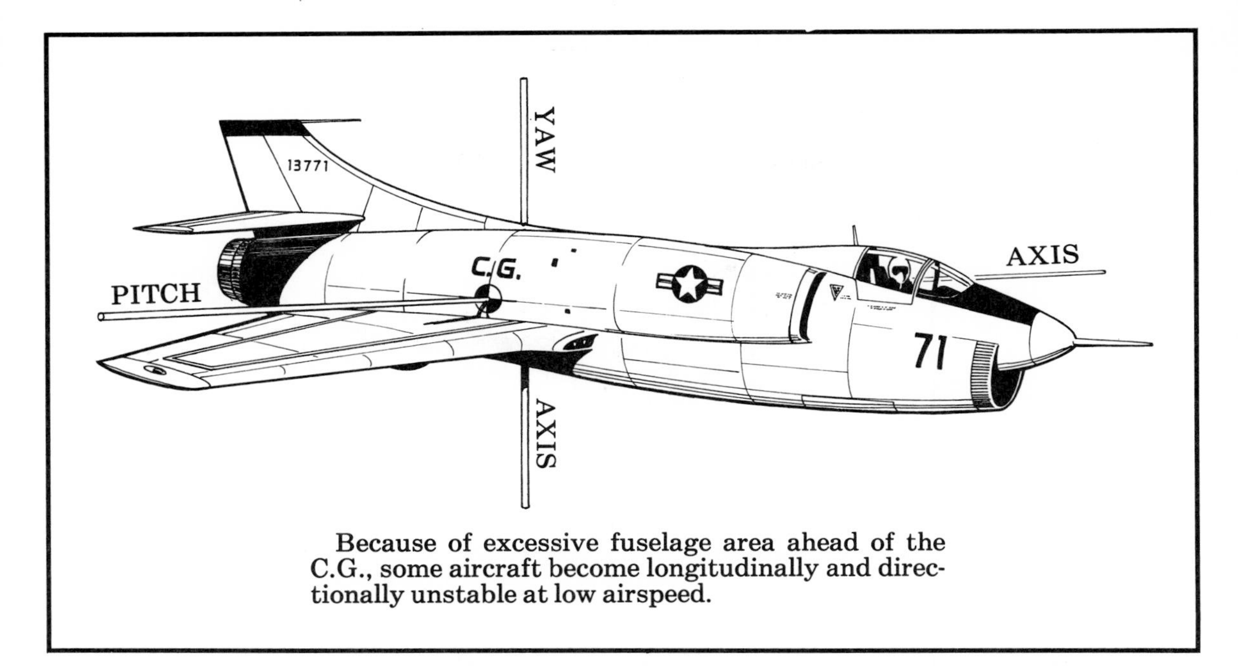

FIGURE 3.7 Longitudinal and directional instability resulting from excessive forward area.

I flew the airplane back to the base and landed. I informed the test engineer that I would not stall the airplane unless the stability was improved.

A few days later the task of investigating the stall of this airplane was given to another test pilot. He lost control of the airplane during the stall and it entered a spin. He stayed with the airplane, trying to recover, until it crashed.

The Effect of Center of Gravity and Weight Distribution The weight distribution relative to the center of gravity will influence the character of the spin. Refer to Figure 3.8. The engine is a considerable distance forward of the center of gravity. A battery is located some distance aft to balance the effect of the heavy engine. Because the arc of rotation of these two objects is far from the c.g., the effect of centrifugal force is more pronounced than if they were located closer to the c.g. During a spin, centrifugal force tends to pull these two objects outward, which results in raising the nose, lowering the tail, and flattening the spin.

The location of the c.g. is important in regard to spin characteristics. At a forward c.g. an airplane is more resistant to spin entries and usually, but

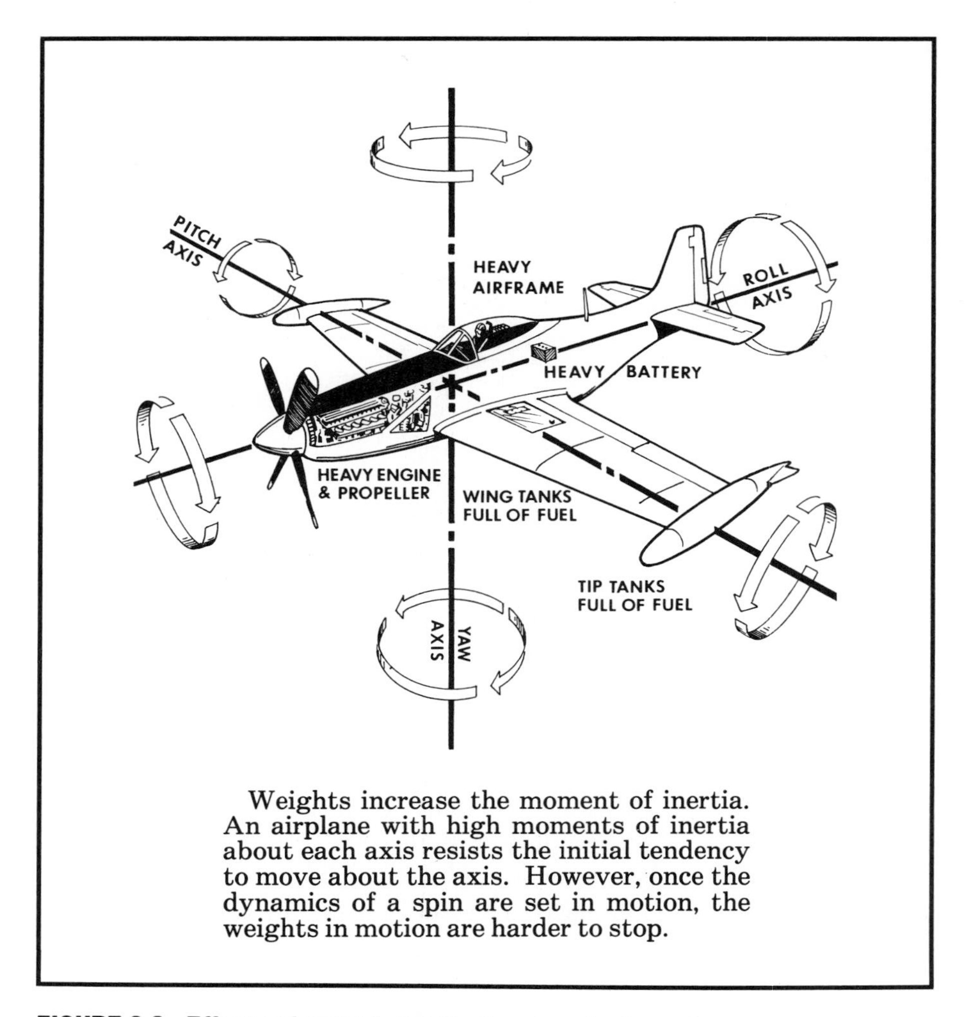

FIGURE 3.8 Effects of weight distribution on spin dynamics.

not always, more reluctant to continue rotation. Some airplanes have worse spin and recovery characteristics at a forward c.g. during a developed spin. However, *all* airplanes are more resistant to the initial spin entry, and as a general rule, a forward c.g. is a safer condition for spins than an aft c.g.

An aft c.g. is certainly conducive to easy spin entries. The nose-down pitching moment is less, and it is far easier to obtain a complete stalling angle of attack. Not only does this aft weight tend to raise the nose and flatten the spin, but the damping arm is shorter and the amount of fuselage surface available for damping is less. In addition, the destabilizing lever arm forward of the c.g. becomes longer as well as adding more surface to contribute to the destabilizing effect. It can be seen that if the c.g. was far enough aft, there would be more area ahead of the c.g. than aft of it. This would have the same effect of shooting an arrow backward; it would swap ends.

Rear-End Design

The shape of the aft fuselage and the tail is an important factor to consider in regard to good spin and recovery characteristics. At one time, those involved in spin research placed considerable emphasis upon the shape of the aft fuselage and the tail in the attempt to obtain ideal spin and recovery characteristics. However, contemporary research has proved that no single factor determines how an airplane will spin or recover. Just because an airplane has an ideally shaped tail or an ideal mass balance does not mean that it will recover from a spin. All factors must be considered together, including the final spin-test results, before determining whether or not the airplane has good spin characteristics.[3]

However, a powerful, effective rudder is very important to spin recoveries. The distance of the tail from the c.g. will determine how much of a lever arm is available to increase the effectiveness of the damping and recovery abilities of the tail. The vertical fin and rudder should be in as effective an airflow as possible at spinning angles of attack. The unshielded-rudder volume coefficient (URVC) is obtained by multiplying the unshielded rudder area by its moment arm from the center of gravity. See Figure 3.9. During a spin, the wing, the fuselage, and the horizontal stabilizer can shield the vertical fin and rudder in such a way that the fin and rudder provide inadequate resistance to autorotation.

Although some T-tail airplanes have excellent spin characteristics, T tails have an adverse effect on the stall characteristics of other airplanes because of the strong downflow from the wings which is encountered at high angles of attack. This sometimes results in an uncontrollable pitch up to an extremely high angle of attack and consequently a very deep stall. The airplane can stabilize in this attitude, and the airspeed is so slow that the controls become ineffective in producing a recovery. T-tail aircraft are also prone to poststall gyrations from this effect.

[3]A document that should be read by serious students of spins is AIAA Paper 80-1580, "Overview of Stall/Spin Technology," by Joseph R. Chambers. It can be purchased from the National Technical Information Service, 5285 Port Royal Road, Springfield, VA 22161.

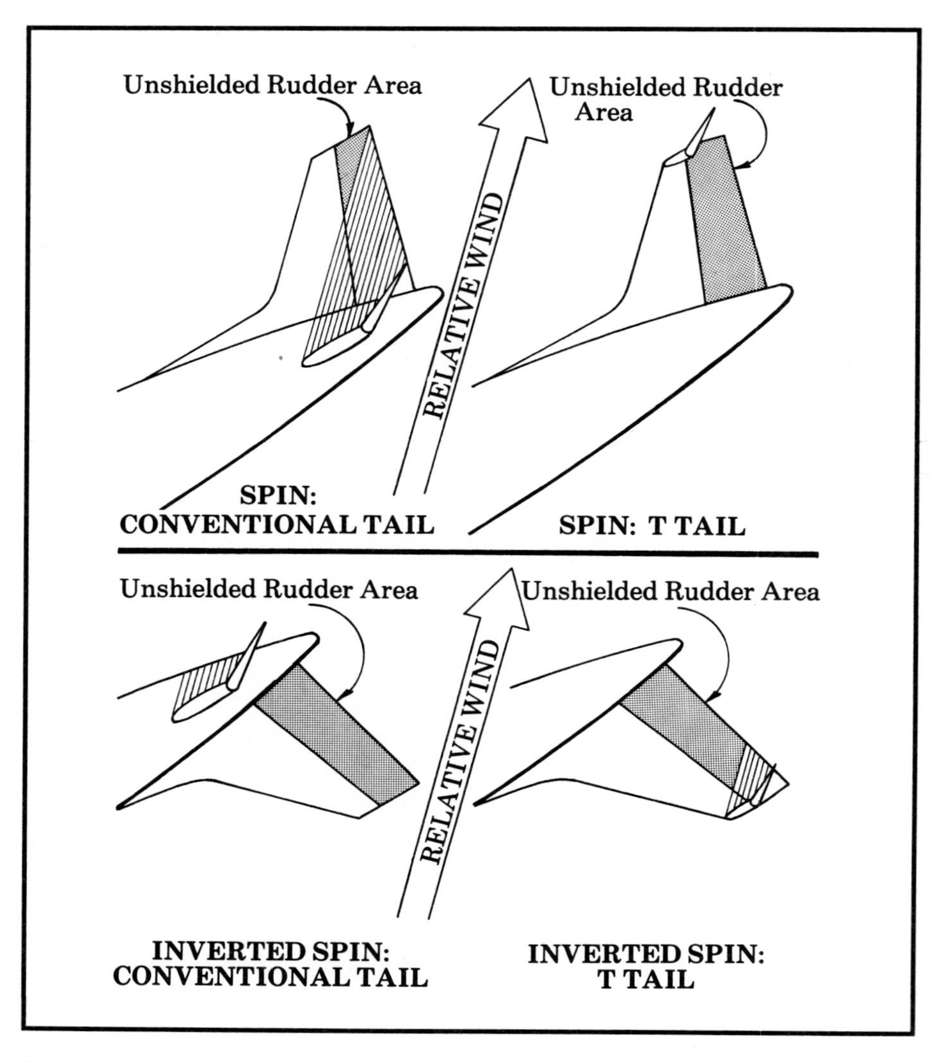

FIGURE 3.9 Unshielded rudder area during spins.

SUMMARY

1. A spin is an aggravated stall that has developed autorotational forces. Influenced by gravity, it descends rapidly along a helical path.
2. The average stalling angle of attack is about 20°. However, the wing reaches a much higher angle during spins. The angle of attack between the two wings will vary. At a 45° spin angle of attack and a rotational speed of about 180° per second, the down-going wing will have an angle

of attack of about 75°. The upgoing wing will have an angle of attack of about 35°. The angle of attack will vary along the wing, with the extreme angles at the wing tips.

3. It is sometimes difficult to tell the difference between a spiral and a spin in airplanes that are highly spin-resistant. As a rule, however, a spiral will display an increase in airspeed, while the airspeed during a spin will remain near or slightly above the stall speed.
4. The three phases of a spin are incipient spin, developed spin, and recovery.
5. Spins, like stalls, can be entered from any attitude.
6. Snap rolls and spins are related.
7. Spins are most easily entered at speeds just above the normal 1-g stall speed.
8. Proficiency in flying an airplane just above the normal 1-g stall speed is extremely important.
9. Yaw control is extremely important in the avoidance of spins.
10. The greatest altitude loss occurs during the first turn of a spin. The average lightplane loses between 250 and 300 ft per turn during a developed spin.
11. There is very little stress being imposed upon the structure during a spin.
12. The use of recovery rudder *followed by* forward stick/wheel is an important sequence for efficient spin recoveries and the avoidance of dynamics that prolong the recovery.
13. An airplane which is very stable about all three axes is likely to be more spin-resistant.
14. Density altitude affects rotational speed. The higher the density altitude, the faster the rotation. Altitude loss is also greater.
15. Spin entries are easier at aft c.g.
16. Most airplanes are capable of spinning when the conditions are right.
17. Inadvertent spins usually occur because of misuse of the controls, or failure to correct for the effects of spiral slipstream, torque, or asymmetric propeller loading.
18. The improper use of aileron is also a spin exciter and a deteriorator of spin characteristics.
19. Holding the aileron opposite to spin rotation can flatten the spin.

20. The use of aileron is usually not recommended during spin recoveries. However, it may be essential to the recovery of some aircraft, for example, airplanes with considerable fuselage mass, or to recovering from flat spins.
21. Some engines may tend to quit owing to the centrifugal loads acting upon the fuel in the carburetor and resulting in poor carburetion. The use of a slight amount of power will keep the engine running, but too much will cause the spin characteristics to deteriorate.
22. The propeller is a gyroscope, and precessional forces can influence spin characteristics.
23. No single factor will determine how an airplane will spin and recover. All factors must be considered together.
24. The total mass of the airplane relative to the density of the air that it is spinning in strongly influences spin characteristics.
25. The spinning mass of an airplane behaves like a gyroscope. This gyroscopic force and its precessional effects will influence spin characteristics.
26. The way the mass is distributed along the fuselage and wings will determine which controls are most effective during recoveries.
27. Too much aerodynamic area forward of the c.g. can deteriorate longitudinal and directional stability and produce unrecoverable spins.
28. The distance from the c.g. of weighted objects can influence the character of a spin.
29. The shape of the aft fuselage and tail is important in regard to spins, but is not the only factor determining good spin and recovery characteristics.
30. A powerful rudder is always an asset to spin recovery.

4

"No one has ever collided with the sky."

-P. T. Barnstorm

LOOKING OUT FOR NUMBER ONE

Using it up
At angles rare
Skimming the earth
With fright and scare
I wonder why
With plenty there
I am skimping
With God's free air?

When assigned the task of discovering the spin characteristics of a jet fighter-trainer, I took it to 35,000 ft, an altitude that I was sure would be sufficient to encounter, recover, or eject from unexpected difficulty. However, I soon discovered that the airplane spun slower and recovered faster at a lower altitude. At high density altitudes spin rotational speeds are faster and altitude loss per revolution is greater. Spin characteristics are usually better and recoveries more positive at low density altitudes.

Obviously, good judgment favors plenty of altitude for spin practice. There is little satisfaction in knowing that an airplane will display its best recovery characteristics just before it drills a hole into the ground.

HOW UP IS HIGH? Opinions vary among instructors concerning how much altitude is necessary for a safe spin lesson. Much will depend upon the type of airplane being spun. Generally speaking, the heavier the wing loading, the higher the entry altitude. If you were spinning a military jet, you'd probably want at least 30,000 ft. That's no problem for military aircraft which gain high altitudes easily.

During World War II, I agreed to work with a cadet who seemed to be making normal progress in every area of his flying except spins. His previous instructor told me that he was deathly afraid of them and would panic every time they tried to perform one.

He was a large man, very well built, with broad shoulders and two of the largest hands I have ever seen.

I took him quietly aside, and after talking to him about his family and other things unrelated to flying in order to put him at ease, I began to talk about spinning the Stearman. I assured him that spinning the PT-17 was perfectly safe. I told him that the airplane would recover from a spin by itself if he would simply let go of the controls.

His flying was reasonably good as we climbed to 3000 ft above the ground—an altitude I considered to be sufficiently high for spinning a Stearman. I kept my voice calm and assured him that he was in for an enjoyable experience. I talked him into a right spin entry and calmly talked to him during the spin.

"Now see, this isn't so bad . . . we've completed one turn . . . now two. . . . Now if you'll just push against the left rudder you'll notice the rotational speed will decrease. . . .

"That's it, push on the left rudder now. . . .

"Push on the left rudder.

"C'mon friend, push on the left rudder!"

We had spun several turns and were throwing away altitude. I decided it was time for me to take the controls, but I couldn't budge them!

I looked into the rearview mirror and continued to force calmness into my voice. There was a look of stark panic written into his expression. Behind his glazed eyes was a barrier to reason.

I used all of my strength to move the stick and rudder. It was as if they were anchored in cement! The whirling landscape seemed to be much closer to my windshield. I stomped hard against the left rudder, hammering against it with my foot. Finally it moved slightly. Then I pushed with adrenaline-boosted strength. The left rudder pedal moved all the way to the stop. The spin rotation stopped just as I was able to shove the stick forward by using the full force of both hands.

I kept the stick and rudder moving to prevent those powerful muscles from again locking up the controls. I stirred the stick around the cockpit until I saw the cadet's hands grasp the rim of the cockpit. He shook his head from side to side as his panic subsided and he returned to a normal state. We were well below 500 ft when I leveled the airplane and started to climb. As I flew back to the airport, I felt sorry for this man. The Lord had undoubtedly created him to become something other than an aviator. He was certainly a misfit in an airplane.

ALTITUDE: MONEY IN THE BANK When you put money into savings, you never know what you might need it for, but it will be available if you need it. So it is with altitude. Six thousand feet above the ground will provide plenty of reserve for spin training in the average light training airplane. However, if you're flying with an instructor who prefers to spin somewhat lower, do not be alarmed. Ordinarily, 3000 ft above the ground is plenty of altitude in a lightplane. I have found that 4000 ft is about right for me. However, if a student shows up with muscular arms and big hands, I'm tempted to add another 1000 ft for each of his fingers.

The altitude loss per turn of spin will vary with each airplane, with the rotational speed, and with the density altitude. The average lightplane will lose between 200 and 300 ft per revolution. Most airplanes will lose less altitude per turn when they are spinning rapidly.

MAXIMUM ATTENTION TO MANEUVERING ALUMINUM Perhaps the description of the composition of our modern contaminated atmosphere should be revised to include aluminum, steel, fuel, and other aircraft components. Some of our busier traffic areas are so cluttered with aircraft that they appear to be airborne junkyards. Obviously, performing aerobatics, including spins, in such surroundings is extremely hazardous. It is the responsibility of those performing aerobatics to make sure that their maneuvers will not conflict with other traffic.

Other traffic is hard to see even under ideal conditions. Unfortunately, some of our better aerobatic airplanes provide less than ideal conditions. For example, sitting in the front seat of an S-2 Pitts is the pits. It's like trying to see the world from a gopher hole. The aft seat of the Decathlon is also bad. Both are exceptionally blind. However, in spite of the propensity of manufacturers to build their cockpits like closets, it is the pilot's responsibility to be sure that the area is clear before performing aerobatics.

AEROBATIC AIRSPACE The regulations pertaining to legal airspace for aerobatics are as follows:

> No person may operate an aircraft in acrobatic flight—
>
> (a) Over any congested area of a city, town or settlement;
> (b) Over an open air assembly of persons;
> (c) Within a control zone or Federal airway;
> (d) Below an altitude of 1500 feet above the surface; or
> (e) When flight visibility is less than three miles.[1]

FORESEEING A FIZZLING FAN Some engines develop mechanical mulligrubs during spins and tend to stop running. Usually a slight amount of throttle will settle their autorotative indigestion and keep them burped and able. Remember that too much power will increase the speed of rotation and make the spin more difficult to stop.

During the years before self-starters, pilots performed their spins within gliding distance of an airport or some other field suitable for dead-stick landings.

Once the propeller stops turning, if the airplane is equipped with a starter, the pilot should use it for restarting rather than resorting to an altitude-consuming dive to gain "windmilling" speed to turn the propeller.

If the airplane is not equipped with a starter, the pilot will have to determine whether it is preferable to throw away altitude in an attempt to restart by diving or to use the available altitude to glide to a suitable landing area.

Some propeller-engine combinations permit easy windmilling and consequent easy restarts. The decision as to what course of action to take will depend upon the airplane, the altitude available, and the skill of the pilot.

Always having a forced-landing area in mind is good insurance, not only while performing spins, but at all other times as well.

THE BODY BRAKE Thinking about handling potential emergencies is a mark of professionalism. If you haven't thought about how you are going to escape from a disabled aircraft, you're likely to succumb to panic when everything turns to worms.

There is very little risk in spinning while under the supervision of qualified flight instructors in airplanes that are approved for spins. However,

[1]FAR 91.71, "Acrobatic Flight."

because I'm in the habit of wearing a parachute while doing aerobatics, I feel much better with one on. The new lightweight parachutes add very little to the gross weight of the airplane and are so comfortable that, if they are available, there is little reason not to wear one. Midair collisions concern me far more than not being able to recover from a spin. However, regardless of the reason for wanting to bail out, you can do it a lot better with a parachute than without one!

The regulations concerning the use of parachutes are stated in FAR 91:15. Parachutes are not required for performing spins with a certified flight instructor or when practicing them solo. However, if you take someone other than a flight instructor with you while performing spins, all occupants of the aircraft must wear parachutes.

Take time to adjust the parachute harness properly. It should be slightly more than snug when you are sitting down and uncomfortably tight when you are standing up. See Figure 4.1.

The D-shaped ring is not the handle to lift the parachute with. If you inadvertently pull it from its pocket, don't just put it back into the pocket without telling someone about it. It needs to be checked to see if the parachute is safe to use.

It is one thing to jump form an airplane while it is being flown straight and level, and quite another thing to bail out while the airplane is floundering toward the ground out of control. Working against the *g* forces can be extremely difficult.

Taking the time to sit in a cockpit, to think through the steps that need to be taken prior to exiting an airplane, and to physically go through the procedure may be the best use of your time you have ever made. I know from experience that emergencies are not conducive to optimum thinking. If you don't know beforehand what procedures you're going to follow, panic will very likely rule the scene.

You should know which openings in the aircraft provide the best exit. Is it a door, canopy, or escape hatch? If your escape must be made through a nonjettisonable door, you should know that you'll probably not be able to open it except at a very slow speed. Pushing the rudder on the side that the door is on will tend to relieve the air loads against it somewhat, but sooner or later you're going to have to take your foot off the rudder. You might have to batter your way through a panel of Plexiglas to provide an escape route. What will you use as a battering ram? Probably your fist. One pilot battered his way through the windshield and exited forward when the wings folded around the cockpit area, blocking his exit through the door and side windows.

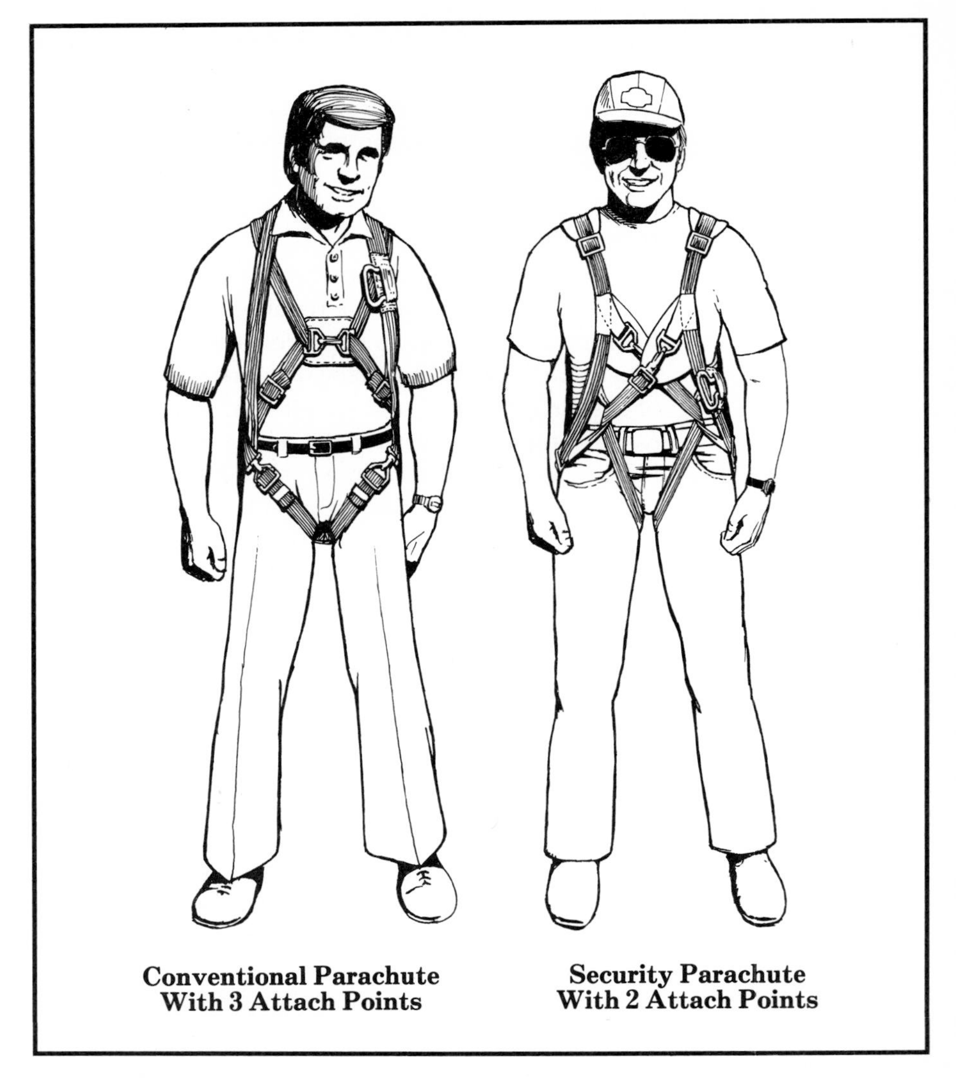

FIGURE 4.1 Two types of parachute harnesses.

You should know the door-jettisoning procedures by heart if the airplane is equipped with a jettisonable door. Get someone to hold the door for you while you practice releasing it.

Open-cockpit airplanes allow the easiest escape. If you happen to be inverted, it is just a matter of releasing your headset and belt, and falling free. If you are right side up, the best way is to stand up in the cockpit and dive over the side.

You should memorize the procedure for freeing yourself from the seat. Will you remove your headset first, then the belt and harness? You need to be careful not to inadvertently unsnap your parachute harness while unfastening belts.

You'll have no difficulty finding the D ring once you're free of the aircraft, so don't grasp it before you leave the airplane. You might accidentally pull it and leave the canopy in the airplane to catch on something. Or you might just flip the D ring out of its pocket and make it difficult to find after you leave the airplane. Keep your hands free of the D ring until *after* you exit the airplane.

Once you're free of the airplane, don't count to three, don't waste time. Look at the D ring, grasp it with both hands, and give it a firm, quick jerk. It will take between 10 and 20 lb of pull to get the job done. Pull it as far as you can, and then throw it away. You can expect a delay of 2 or 3 seconds before the parachute opens and you are jerked upright. There tends to be a hypnosis that makes it seem as if there is no hurry in getting to the ripcord. There is no sensation of falling, no feeling of ill-being. However, you will fall about 1000 ft in the first 10 seconds and 1000 ft for each succeeding 5 seconds.

Be determined to get to the rip cord as soon as you're out of the airplane. By the time you get to it, you will be well clear.

During the last 50 ft of descent, the ground will seem to come toward you pretty fast. Put your legs and feet together and bend your knees slightly. Tuck your head down against your chest. It takes practice to land lightly on your feet. Don't try it. Just roll with the punches.

If you see that you're going to land in trees or power lines, cross your feet and keep your knees bent. If you are landing in trees, put your hands under your armpits. This combination will tend to protect some critical arteries in event of a severe puncture wound.

If you're going to land in water, be prepared to slip out of your harness as soon as you touch the water. Start well ahead of time to get ready. First, be sure you're sitting well back in the harness. If you're in a conventional three-snap harness, unsnap the chest strap and place your hands on the leg strap releases. When you hit the water, release them.

Once you're free of the harness, you'll be free to swim or to fight the sharks. With luck, a friendly boat will be coming your way.

There has been much discussion of how much altitude you need to make a safe bailout. However, if there is no hope of controlling the aircraft, disregard any thought of altitude and work as rapidly and as efficiently as possible.

One thousand feet is usually thought of as the lowest possible altitude to attempt a bailout. However, much will depend on the speed and attitude of the airplane at the time. Successful bailouts have been accomplished at much lower altitudes.

Of course, aircraft equipped with ejection seats greatly simplify the problems involved in leaving an airplane and greatly increase the chances of survival. Most modern ejection seats have zero altitude capabilities if the airplane is upright at the time of pilot ejection.

SUMMARY

1. Spin characteristics are usually better, and recoveries are better, at low density altitudes.
2. Good judgment favors plenty of altitude for spin practice. There is little satisfaction in knowing that an airplane will display its best recovery characteristics just before it drills a hole into the ground.
3. Altitude is like money in the bank; it's nice to have when you really need it.
4. Aircraft with heavy wing loadings need more altitude for spin practice than airplanes with light wing loadings. Four thousand feet is ordinarily enough altitude for the average lightplane.
5. The average lightplane will lose between 200 and 300 ft per revolution in a developed spin. It will probably use considerably more during the first turn.
6. It is the responsibility of those performing aerobatics to make sure that their maneuvers will not conflict with other traffic.
7. You should be aware that non-fuel-injected engines may tend to stop running during a spin. You should know your air-start procedures and have a forced landing area in mind. Carrying a slight amount of power during a spin will assure that the engine will keep running. However, power is usually detrimental to spin characteristics and too much power should not be used.
8. The windmilling air start after the propeller has stopped turning usually consumes a lot of altitude. If the airplane is equipped with a starter, use it.
9. It is a mark of professionalism to think about handling potential emergencies.

10. Parachutes are not required for performing spins with a certified flight instructor or when practicing them solo. However, if you take someone with you other than a flight instructor, all occupants of the aircraft must wear parachutes.

11. It reflects good judgment to wear a parachute during any aerobatic practice.

12. Sit in the cockpit and practice emergency egress procedures. Know how you will get out of a particular aircraft before you need to. It is good insurance!

13. When bailing out, you will fall about 1000 ft in the first 10 seconds and 1000 ft for each 5 seconds thereafter.

14. Do not place your hand on the D ring before leaving the airplane. You will have no trouble locating it after you're free of the airplane.

15. As soon as you are out of the airplane and have placed your hand on the D ring, it is time to pull it. You will be clear of the airplane.

16. During the last 50 ft you will realize your fast descent rate. Put your legs and feet together and bend your knees slightly. Tuck your head down against your chest. Don't try to land on your feet, just roll with the punches.

17. If you see that you're going to land in trees or power lines, cross your feet and bend your knees. Put your hands under your armpits to protect critical arteries.

18. If you are landing in water, be prepared to slip out of your harness as soon as you enter the water.

19. Don't think about altitude loss when fighting your way out of an uncontrollable airplane. Work as rapidly and efficiently as possible.

"Maneuver softly and avoid a rough stick."

-P. T. Barnstorm

STICK-AND-RUDDER ANTICS

What the pilot does with the controls will not only influence a spin entry but will often influence the character of the spin. Frequently the control positions that excite inadvertent spins will also result in spins that are difficult to recover from in some airplanes.

For example, one of the easiest ways to enter an accidental spin is through a skidding turn. Once the bank is established, if you're in the habit of influencing the rate of turn by use of the rudder, you may someday be surprised by an unwanted gyration.

THE HEALTHY TURN

All healthy turns are ball-centered, neither slipping nor skidding. Of course, some rudder may be necessary to prevent unwanted yaw caused by power, or propeller prescessional forces. However, the rudder's primary purpose is to correct for adverse yaw resulting from the unequal drag of the ailerons and for the asymmetric lift and drag resulting from the variations of the angle of attack of the two wings. See Figure 5.1.

The turn rate should always be controlled by varying the bank, never by holding the rudder into or against the turn.

A SPIN TRAP

Let's consider a typical situation from which an accidental spin can occur. A pilot who has developed the habit of skidding through turns becomes distracted while turning left from the base leg to the final approach. The pilot permits the airspeed to become too slow while holding left rudder and thus skids through the turn. The use of excessive rudder results in an overbanking tendency. This requires the use of right aileron. As the bank steepens, left rudder also yaws the nose below the horizon. The use of an abnormal amount

FIGURE 5.1 Aerodynamic forces causing adverse yaw in airplane while banking.

of up elevator is required to hold the nose in the desired attitude. The angle of attack of the wing is now greater than normal for the angle of bank. Finally, the left wing stalls because of low speed and too great an angle of attack. (See Figure 2.21.) The depressed aileron also contributes to inducing a stall. Its drag, plus the left yaw caused by holding left rudder, results in a quick roll-off to the left as a spin is entered. There is seldom any warning from prestall buffeting, and the first indication the pilot has of trouble is a quick spin entry.

Pilots who are skilled in spin recoveries, and have been exposed to spin-inducing situations at safe altitudes, will probably never enter an inadvertent spin. However, if they should, their recoveries would probably be instant and automatic, with a minimal loss of altitude.

Again, let's consider what may occur to the unskilled pilot who has skidded all the way into a spin. The pilot pulls the stick all the way aft in an attempt to pull the nose up and uses full opposite aileron in an attempt to stop the rotation. The use of full up elevator holds the wings at a high angle of attack, and the use of aileron opposite to rotation effectively increases the angle of attack on the stalled wing, increasing the drag and further aggravating the autorotation. One of the easiest ways to flatten a spin and make it unrecoverable in some airplanes is to use aileron opposite to spin rotation.

PRO-SPIN AILERON The use of aileron in the direction of the spin (right spin, right aileron) will usually increase the speed of rotation. As a result, it may take a little longer to stop the rotation during the recovery. However, because pro-spin aileron tends to develop a healthy amount of roll relative to yaw, the spin is less likely to flatten than if aileron opposite rotation is used.

AILERON USE IN FLAT-SPIN RECOVERY Because the use of aileron into the spin direction increases rotational speed, it is possible for the centrifugal force to become so strong that it would tend to flatten the spin. However, one of the recommended ways of recovering from flat spins in some airplanes is by using fully deflected aileron in the direction of the spin. This procedure utilizes the drag of the depressed aileron to resist autorotation and to lift the outside wing. The lifting tends to recouple roll with yaw and to restore a normal spin pattern. Also, the raised aileron on the inside wing will serve as a spoiler to the lift and autorotational force being contributed by this retreating wing. See Figures 5.2*a* and 5.2*b*.

There may be a temporary increase in rotational speed as the yawing energy translates into rolling energy. However, as the nose again drops to a normal spin attitude, a spin recovery can be effected.

Normal spins incorporate both yaw and roll. When the roll portion translates into pure yaw, a flat spin pattern usually develops. When this occurs, a recovery may be difficult or impossible.

In most cases, the use of aileron with or against the spin is not recommended. However, certain highly spin-resistant airplanes may require the use of opposite aileron to enter and, in some cases, to continue the spin. However, unless the pilot's operating manual *specifically recommends* this procedure, don't experiment with it on your own.

The ailerons on most airplanes will tend to "float" into the direction of the spin. Pressure against the ailerons will be necessary to hold them centered. Use whatever opposite pressure is necessary to keep the stick or wheel centered.

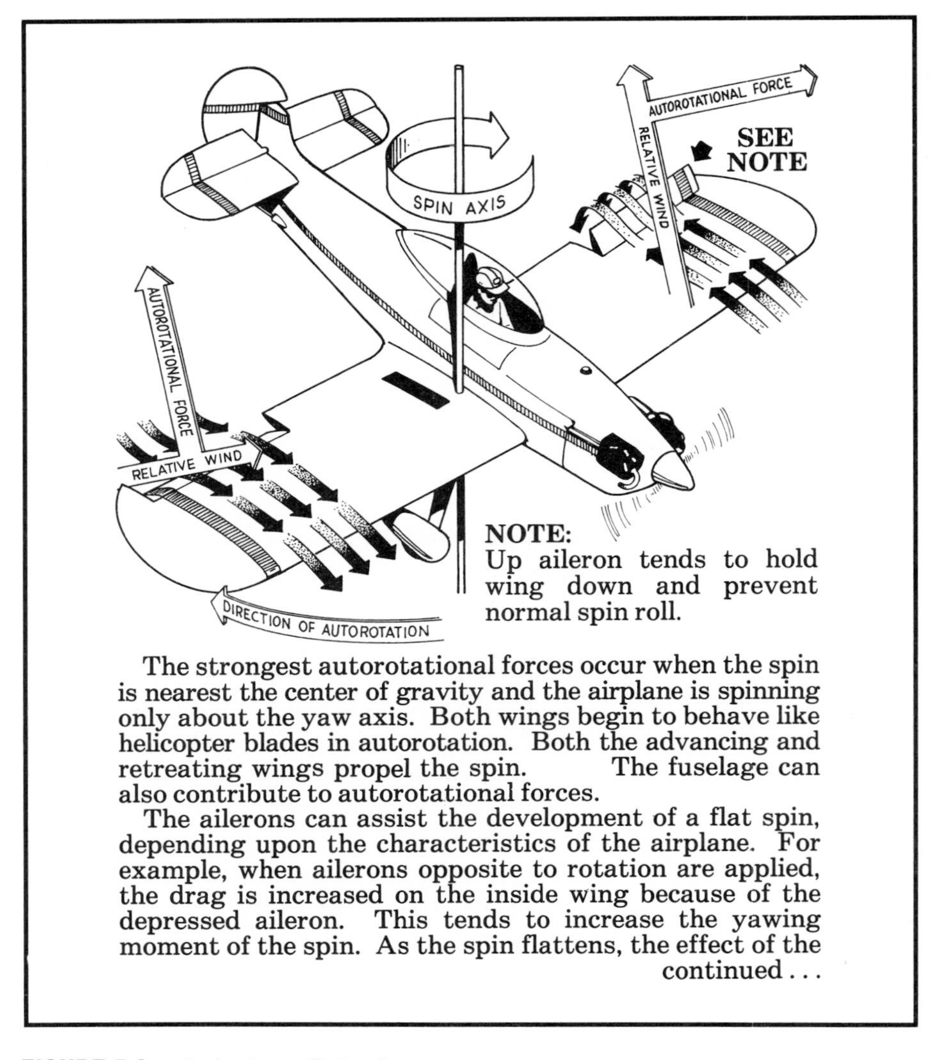

FIGURE 5.2*a* Inducing a flat spin.

PRACTICE SPINS Although spin entries should be practiced from situations that normally produce inadvertent spins, introductory spins are usually entered from normal 1-g stall attitudes. An introductory practice spin would proceed in the following manner.

Climb to a safe altitude and make whatever clearing turns are necessary to be sure that a spin will not conflict with other traffic. Align the airplane with the section lines or a distant object. Throttle back to a low power setting

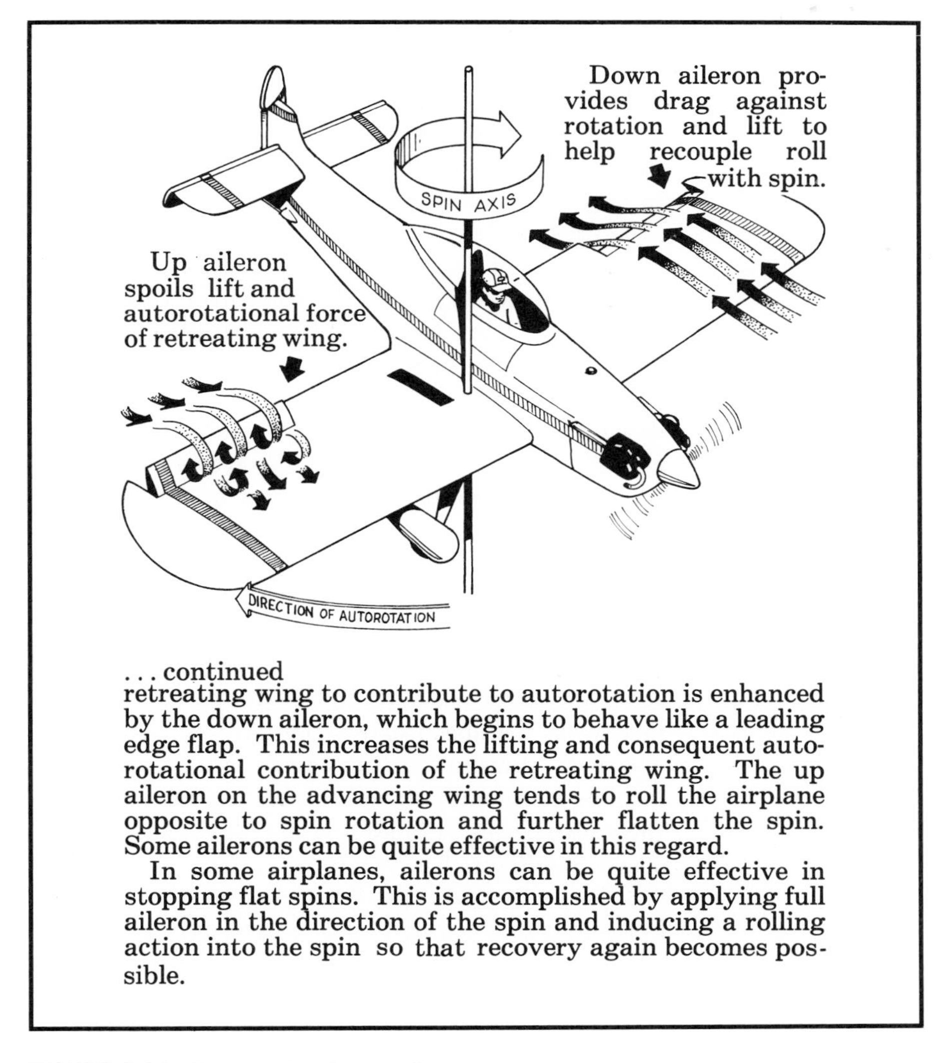

. . . continued
retreating wing to contribute to autorotation is enhanced by the down aileron, which begins to behave like a leading edge flap. This increases the lifting and consequent autorotational contribution of the retreating wing. The up aileron on the advancing wing tends to roll the airplane opposite to spin rotation and further flatten the spin. Some ailerons can be quite effective in this regard.

In some airplanes, ailerons can be quite effective in stopping flat spins. This is accomplished by applying full aileron in the direction of the spin and inducing a rolling action into the spin so that recovery again becomes possible.

FIGURE 5.2*b* Recovering from a flat spin.

while increasing the angle of attack. Gradually pull the nose above the horizon until the airplane is slightly above landing attitude.

About 5 miles per hour above the normal stall speed, apply full rudder in the direction of desired spin. If a left spin is desired, apply full left rudder. As the airplane yaws to the left, the left wing will stall and the airplane will yaw and roll to the left. As this occurs, apply full up elevator, being careful not to use aileron in either direction. As soon as autorotation develops, pull the throttle back to idle, or nearly so. A slight amount of power may be desirable to keep the engine running on some airplanes. (Remember that power tends to increase rotational speed; use it sparingly.)

Hold the rudder and elevators fully deflected for as long as autorotation is desired. Most airplanes will reach the maximum rotational speed after four to six turns. However, there is little point in letting the airplane spin beyond two turns. Most airplanes will be spinning adequately enough by two turns to provide a good learning situation for practicing a recovery.

Use the method recommended in the pilot's operating manual, which usually consists of first using full rudder against the spin rotation, followed by the use of forward stick/wheel to neutral. However, during spin practice the pilot should learn to observe the airplane's reaction to recovery rudder. It will indicate just how much forward stick/wheel is needed to complete the recovery. If the airplane responds slowly to the rudder, or not at all, full down elevator may be necessary to stop the spin. If the rate of rotation greatly decreases with the use of recovery rudder, moving the stick/wheel forward to neutral may be all that is necessary.

AN IMPORTANT SEQUENCE

The rudder-followed-by-stick sequence is important. Premature use of down elevator tends to deflect airflow away from the vertical fin and rudder, making these surfaces ineffective in damping and stopping rotation. Also, premature use of down elevator may result in translating yaw into roll, increasing the speed of rotation. When this occurs, the rotational radius decreases, transmitting the energy into a smaller circle. An ice skater performing a pirouette utilizes this same principle. By pulling the arms inward, the skater moves the mass to a smaller circle and thereby increases rotational speed. See Figure 5.3.

It is easy to make the transition into an inverted spin from this situation. For example, if forward stick is used prior to the rudder and the pirouette effect develops, the airplane may already be pointed well down and will invert when the opposite rudder is used. What was intended as recovery rudder now has the potential to hold the airplane in an inverted rotation while

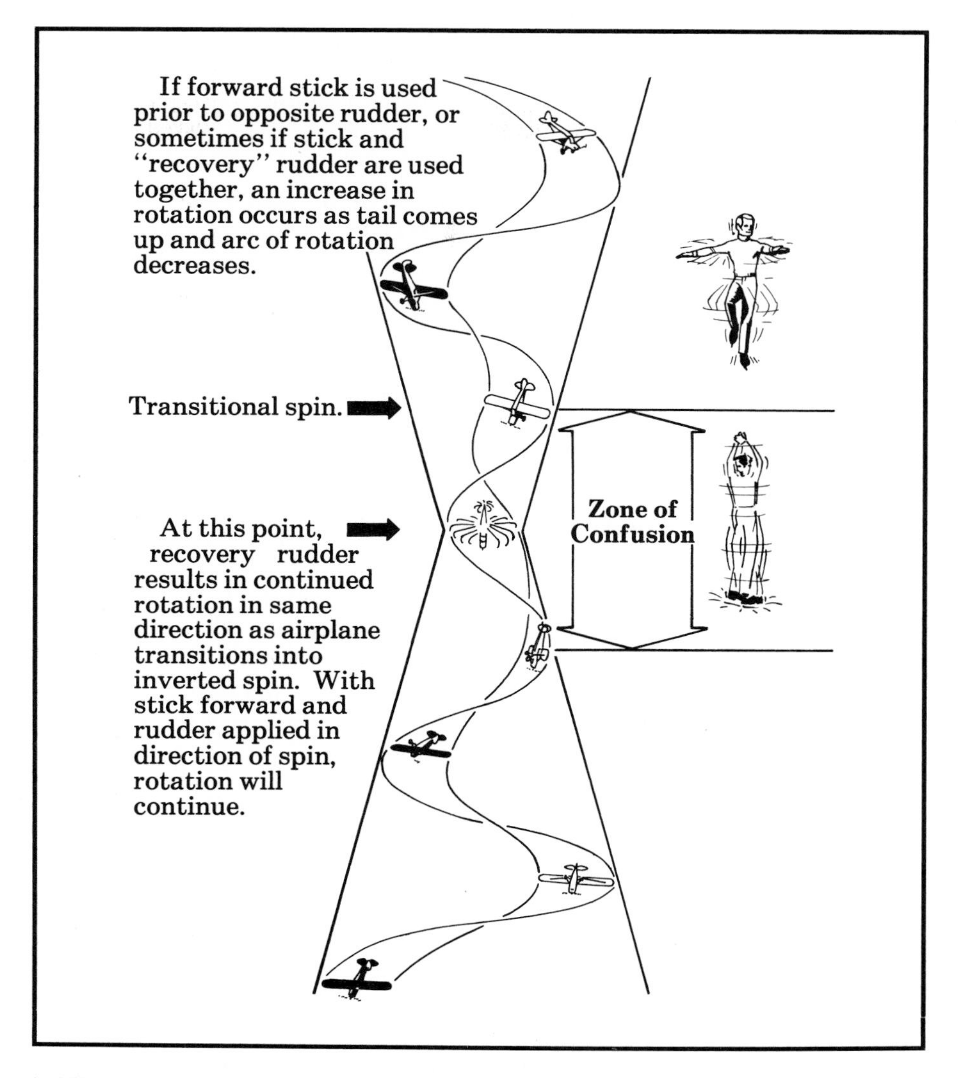

FIGURE 5.3 Transition from erect to inverted spin.

essentially rotating in the same direction. This transition is insidious and often unrecognizable except to those who are experienced in aerobatics.

Spin recovery procedures can vary considerably from airplane to airplane. The statements I've been making apply *generally* to most airplanes. However, there are times when seemingly unorthodox methods must be used. For example, the use of aileron during a spin recovery is normally not recommended, but in a few exceptions they can be used against the spin to expedite a recovery. As previously mentioned, ailerons used in the direction of the spin are essential to stopping the spin of airplanes with heavily loaded fuselages, such as the T-38.

PRACTICE, PERCEPTION, AND REACTION The Aeronca C-3 was a popular lightplane during the 1930s. It performed remarkably well with a 2-cylinder, 36-hp Aeronca engine. It was light and strong and was maneuverable enough to find its way into the spotlight in many air shows of that era. Among other maneuvers, I used to perform two turns of a spin from 500 ft to a landing. It was an airplane that would enter a spin easily, but its recovery characteristics were quick and positive.

I had been experimenting with whip stalls entered from ground level. I would zoom into a vertical climb, permit the airspeed to drop to zero, slide backward slightly, and then let the airplane whip straight down.

I had practiced the maneuver several times, and there was always sufficient altitude to recover. However, it was a warm day and the density altitude was undoubtedly higher than what I was used to. As the nose whipped down, the proximity of the ground, the rate of whip, and the dead feeling of the controls forecast an impending fiasco.

I jammed the throttle all the way forward and tried to flatten the C-3 to a level attitude without encountering another stall. With the ground threatening to come through the windshield, I had no choice but to attempt to pull the nose up—stall or not. As I pulled against the stick, one wing stalled, threatening a spin. I pushed opposite rudder, careful not to use aileron. Suddenly, the other wing dropped. Again I reacted with opposite rudder. The nose was almost level, but I was settling rapidly when I hit the ground hard!

The right landing gear tore loose, swung momentarily from the brake cable, crashed through the cockpit fabric, and battered my right leg. It then fell free of the airplane. The C-3 bounced back into the air, and I was able to keep it there.

After circling the airport several times to regain my composure, I landed on one wheel, the right wing dragging lightly as I lost speed and lateral control. There was no further significant damage, other than a scuffed wing tip.

It was a shoddy demonstration of judgment and flying skill, but I was alive because I did not spin. No matter how many other things pilots may do wrong, if they can instinctively avoid spins, their chances of survival are greatly increased.

Interestingly enough, it isn't the air show performer who is most likely to become a stall/spin statistic. More than likely it will be the pilot who seldom, if ever, exceeds a 30° bank and hasn't practiced a stall, let alone a spin, in years.

Just as physical exercise improves your physical health, proper flying exercises will improve your ability to fly an airplane.

LOGIC AND LONGEVITY Actual spin entries and recoveries practiced at a safe altitude under conditions from which inadvertent spins are most likely to occur are excellent insurance for a long and enjoyable flying career.

Most spins are practiced illogically and do not yield maximum benefit. Most practiced spins are induced after the airplane has penetrated deeply into the stall. Spin entries can be induced more quickly at speeds just above the normal 1-*g* stall speed. A quick asymmetric stall can be obtained while the controls are still effective. If the pilot waits until the airplane is deep into the stall before applying rudder in the direction of the intended spin and pulling the stick fully aft, the controls will not be effective enough to make an immediate and decisive entry. The airplane may actually roll opposite to the direction of the rudder before it finally responds in the direction the pilot intends. An airplane is more responsive to spinning at speeds just above the normal stall speed.

At a safe altitude, pilots should frequently practice flying at speeds just above the stall. They should practice rolling in and out of level, climbing, and descending turns, making them as steep as possible. They will soon recognize that spins are easily induced under these conditions.

While the airplane is being flown at a high angle of attack, an asymmetric stall can be quickly induced by misuse of the controls. For example, when the proper amount of rudder is being used with the ailerons, the proper roll rate response can be expected. However, if inadequate rudder is used, adverse yaw distracts from the roll, causing the pilot to deflect the ailerons more than would normally be required. This increases the drag on the wing with the downward-deflected aileron, and effectively increases the angle of attack on that wing. The other wing, though it may be partially stalled, still has a healthy amount of lift and rolls the airplane into a spin.

Inadvertent yaw may be caused by other factors, such as not correcting for the inherent tendency for the airplane to yaw left at low airspeed and high power. One of the situations which has resulted in many an accidental spin after takeoff is a steep climb in which the pilot attempts to maintain the airplane's heading by use of right aileron only, resulting in further left yaw and an effective higher angle of attack on the left wing. Finally, the left wing stalls, and under the influence of the power-induced left yaw, the airplane quickly enters a spin to the left.

These aileron and power-induced spins can be practiced safely with a competent flight instructor. The practice would proceed in the following manner:

1. Enter a steep climb using maximum allowable power. Do not correct for the left yaw effect that is induced by power.
2. While holding a steep climb attitude, hold a straight flight path by using right aileron and flying right wing low.
3. The left wing will stall first and cause a left roll. The power effect will couple yaw with the roll, and a left spin will develop without the use of left rudder.
4. Recovery should be accomplished within one turn of rotation by closing the throttle, using right rudder to resist rotation, and then using forward stick/wheel. Of course, the ailerons should also be centered.

Some airplanes will not induce enough left yaw to effect a quick and positive spin entry. The instructor can simulate the left yaw effect by applying a small amount of left rudder.

Practice in this spin entry will make the pilot aware of the importance of using sufficient right rudder to correct for the effect of power during a climb.

WHEN THE ENGINE FIZZLES Another typical stall/spin accident that has occurred after takeoff is the result of improper flying technique while attempting to maneuver the airplane to a forced landing after a power failure.

A sudden and complete loss of power after takeoff requires prompt action. The airplane is already in an attitude for a stall, and if the nose isn't lowered immediately to reduce the angle of attack, a stall will develop quickly.

With the speed already dangerously slow, pilots have been known to maneuver excessively in an effort to land in a selected area. Spins can quickly develop at the high angle of attack generated by sudden steep turns and the further loss of flying speed caused by the drag of large control deflections.

Sometimes this excessive maneuvering is the result of not being psychologically prepared for a forced landing.

One thing a pilot should be prepared to accept during a forced landing is damage to the aircraft. Even the most skillful of pilots cannot avoid bending the airplane under certain conditions.

The worst possible mental activity at the moment of engine failure following takeoff is concern over the value of the airplane. To hang the nose of the airplane high upon nothing while spending time looking for a perfect landing area is a futile and often fatal activity.

Without power, gravity is in command of the situation, and unless it is skillfully appeased, it can ruthlessly bring about a fatal conclusion. Gravity *must* be utilized as a source of power to maintain airspeed and control of the airplane. This means that when the engine quits, a pilot's actions will decide how the airplane will descend—controlled or uncontrolled. It *is* going to come down!

The object is survival! The attitude and the rate at which the airplane hits the ground will determine the probability of that result. The accidents that I've seen over the years lead me to conclude that if the average lightplane lands somewhere near its normal landing speed and attitude while under control, regardless of the type of terrain it comes in contact with, the possibility of survival is good.

ATTEMPTING THE IMPOSSIBLE The descent rates resulting from a stall, spin, or excessively low airspeed are lethal, and the potential for survival is greatly decreased. The stall, spin, or rapid descent rates are frequently the result of attempting to "stretch" a glide in an attempt to reach an impossible landing site.

It is always a foregone conclusion that if stretching results in a speed below an airplane's best glide speed, it will fall short, and you might as well resign yourself to landing in whatever terrain exists between you and the previously selected landing area. This may mean a landing in trees, houses, or other obstructions. If you don't attempt to stretch the glide over these objects, but rather head for their bases, putting the fuselage between these solid barriers, you will probably walk away from the result.

Many a spin has started at treetop obstruction level as pilots attempt to clear the trees through the stretch process. Even if you barely brush a wing through a treetop, the resultant yaw can excite autorotation.

It is well to mention here that excessive speed is not the answer. Use the best glide speed for distance, slowing the airplane as much as possible near ground level by extending flaps or by slipping.

THE TURNING-AROUND SYNDROME Attempting a 180° turn to the runway when the engine fails after takeoff has resulted in a great number of pilots becoming stall/spin statistics.

Unless there is a long runway behind you and you *know* without any doubt that you have sufficient altitude to complete the maneuver, turning back to the runway can be an extremely dangerous procedure.

A spin can develop more rapidly out of a turn because of the higher stalling speed, higher angle of attack, and pro-spin control position. Consequently, rolling into what will undoubtedly be a steep turn from low airspeed is an open invitation for a spin. Skidding the turn to hurry the turn rate is also a common and often fatal error under these conditions, as excessive yaw promotes an imbalanced stall with resultant autorotation.

Also, as the airplane turns past 90° and the tail wind increases the ground speed, the pilot often interprets this as airspeed and increases the angle of attack to stretch the glide. This is often an inducement for a stall or spin, or, at the very least, an extremely fast descent rate with considerable forward speed.

If there is any significant wind and any doubt whatsoever about completing a 180° turn, it is far better to land into the wind as much as possible, since contact with the ground or obstacles is made at a greatly reduced speed. Crashing downwind with the additional ground speed caused by the wind may tip the scales in favor of a tombstone.

John Margwarth, a flight safety staff engineer for Lockheed Aircraft, has written a commentary on low-altitude engine failures that I believe is worth quoting here:[1]

Most GA pilots probably have given some thought to the subject of engine failure during the initial climb-out following a takeoff. One pilot estimated that he could make a 180 degree turn after an engine failure, using climb airspeed, and not lose more than about 150 feet in altitude during the 180. The aircraft being discussed is a popular 100-HP high-wing two-seater, grossing 1600 pounds.

Okay, let's look into the matter of complete power failure at low altitude and start with the following data and assumptions for the above mentioned aircraft, flying from an airport having a runway 3900 feet long:

(1) Elevation of airport, 1000 feet; zero wind; temperature approximately 60 degrees F.

(2) Wheels leave the ground with 3000 feet remaining to the end of the runway.

(3) Best angle-of-climb airspeed—70 MPA IAS.

(4) Best rate of climb airspeed—75 mph or 110 ft/sec.

(5) Seconds to end of runway at 110 ft/sec—27.

(6) Rate of climb—600 ft/minute or 10 ft/second.

[1]Used by permission of author. This material also appeared in the April 1977 newsletter of the Lockheed-California Company Pilots Club.

(7) Altitude above ground at end of the runway—270 feet.

(8) CAS stall speed: 55 mph at one G with flaps up, 59 mph in an 30 degree bank, 65 mph in a 45 degree bank and 78 mph in a 60 degree bank.

(9) Best engine-out glide angle in one G wings-level flight, 70 mph IAS, with zero wind—1:9 (actually nearer 1:8.7).

(10) Diameter of turn at 75 mph (110 ft/second) and a bank angle of 45 degrees—approximately 750 feet. (Formula: Radius is ft/sec squared divided by 32.2 and the tangent of the bank angle).

Now let's take a look at how far we fly around a 180 degree arc starting at the end of the runway in a 45 degree banked turn at 75 mph. That distance is approximately 1180 feet (3.14 x D/2). Using a glide ratio of 1 : 9, the loss in altitude around the turn would be 130 feet. This might appear to be pretty good until you consider the following:

(1) In a 45 degree banked turn at 75 mph, 1.4 G, the actual glide path will be much steeper than 1:9, whatever the airspeed.

(2) After completing the 180 at 75 mph the aircraft is displaced 750 feet laterally from the runway.

(3 Because of the preceding item, the equivalent of two more 90 degree turns, or two 45 degree turns with a straight flight path in between, are required to get back over the runway.

(4) After a power loss, the pilot must react promptly to control the airspeed and to guard against excessive bleed-off.

(5) Trees, telephone wires, and altitude allowance for the flare are items to be considered.

As a matter of interest, some informal flight checks on an aircraft configuration approximating the subject aircraft show that the altitude loss during a *180 degree* turn probably will use up *250 feet* if flying at 75 mph and a 45 degree bank angle. In a 60 degree banked turn the altitude loss could be less, except for the fact that stall speed is increased to 78 mph CAS. A couple more observations: with zero wind, standard day or warmer, the climb flight path angle for such an aircraft at 1600 pounds is more shallow (disadvantageous) than the best power-off glide angle. Taking off into a stiff headwind gets you substantially higher at the end of the runway, but after completing a 180, the real estate goes by awfully fast.

One more comment. With this sort of evaluation, a pilot can attempt to figure out where on the downwind leg of a given single-runway airport that the landing direction should or might change after an engine failure in the pattern. Velocity or the wind will effect a proper choice, as well as altitude (AGL) and lateral displacement of the downwind leg. And, in some cases analysis may indicate that the landing should be made away from the runway/airport. Note: Stall, climb and one G wings-level glide data are in close agreement with the aircraft owner's manual.

An excellent practice maneuver that graphically demonstrates some of the hazards involved in a quick 180° steep turn from low airspeed is as follows:

1. At a safe altitude, climb steeply at maximum climb power until the airspeed is within 5 miles per hour of the stall. Note the altitude and close the throttle completely, simulating a complete engine failure.
2. Execute a steep 180° turn. The object is to lose as little altitude as possible. This requires as steep a bank as possible at the lowest possible airspeed.
3. Note the altitude the moment 180° of turn has been completed.

It has been my experience that most pilots enter a spin on the first attempt. However, with a little practice they are able to avoid spinning and complete the maneuver with a loss of altitude of 250 to 300 ft in the average lightplane.

It is a critical maneuver that demands an excellent feel of the airplane and perfect coordination. Otherwise, a spin is likely to result.

The object of this training maneuver is not to encourage you to turn back to the airport in the event of an engine failure after takeoff. On the contrary, it reveals how critical the maneuver is, and will discourage you from flying beyond your proficiency in such a circumstance. It will also help you to develop the skill of flying an airplane on the very edge of a stall or spin during a critical turn.

Years ago I watched as a pilot of a Security Airster deliberately turned off his engine directly over the airport and entered a pattern for a dead stick landing. As the 5-cylinder engine died, the propeller kicked through a final spasm, and then stiffened.

The Security was a low-wing, side-by-side airplane with plenty of drag and a short glide. The pilot had been accustomed to airplanes with a better glide ratio.

As he turned into the final approach it was obvious that he was too low to glide over the high-tension power lines that bordered the airport. He pulled the nose up as he approached the wires, then, finally convinced that he could not clear them, attempted to turn away. The airplane promptly entered a spin, but as it did, one wing snagged a large billboard, which stopped the spin and cushioned the impact. The pilot and his passenger, a friend who was also a pilot, were injured, but after a few weeks in the hospital, they were flying again.

FAA flight tests include departure and arrival stalls, but what about departure and arrival spins? What about the spin that occurs somewhere in the traffic pattern, especially on the final approach to a landing? What are the circumstances that are most likely to generate a catastrophic stall or spin?

TWO HYPOTHETICAL DISASTERS Let's consider a couple of hypothetical situations that have undoubtedly occurred in a similar manner many times over.

A pilot inexperienced in mountain flying descends into a mountain valley with the intention of landing on an airstrip. Its elevation is 8000 ft and the density altitude is 10,000 ft. The wind is blowing in excess of 20 knots, creating considerable turbulence.

As the pilot enters the downwind leg of the pattern, the airplane flies over some low hills that border the length of the airport. Because of the airplane's proximity to the hills, the presence of the tail wind, and the high true speed caused by the altitude, the pilot becomes aware of excessive speed.

Owing to the turbulence and the anxiety of attempting a landing in the mountains, and troubled by the new awareness of speed on the downwind leg, the pilot is unaware that the airspeed is lower than normal. The resultant nose-up attitude and high angle of attack are masked by the mountainous references that surround the airplane. The pilot has been accustomed to a flat and definitive horizon.

The awareness of excessive ground speed leads the pilot to reduce power considerably, dangerously slowing the airspeed. The pilot is unaware that the angle of attack of the wing is reaching a critical value.

Because of the surprising ground speed, the pilot suddenly realizes that the airplane is now a considerable distance downwind of the airstrip and therefore banks steeply to turn into the wind to start the final approach. As the airplane banks into the turn, the pilot experiences stall buffet and immediately reacts by adding full power. At this instant the airplane quickly enters a rapid spin and crashes.

Why did the airplane spin the moment power was applied? Because the nose pitches up in most airplanes when power is applied, increasing the angle of attack. If the pilot does not release the back pressure which is being held against the stick/wheel during a steep turn as power is added, the angle of attack may become critical. However, this pitch-up reaction is masked during a steep turn, and the angle of attack may increase without the pilot becoming aware of it.

Airplanes always react more vigorously to a stall or spin at high density altitudes. If a wing drops mildly during a stall at a density altitude of 2000 or 3000 ft, it will roll out considerably faster at a density altitude of 10,000 ft because the air is thinner and the resistance to rolling is less. Spin rotational speeds can be very rapid at this altitude.

Let's consider another situation from which a spin might occur during final approach.

Suppose that because of poor visibility, a pilot flies a pattern close to the runway to keep it in sight. Because the object is to get the airplane on the ground on the first attempt, the pilot slows the airplane to minimum speed and makes a low approach. During the turn from base to final, the pilot sees that the airplane is overshooting the projected runway centerline. Having been trained to avoid steep turns close to the ground, the pilot consciously or unconsciously pushes rudder to increase the turning rate and uses opposite aileron to prevent the bank from steepening. Suddenly the airplane spins in the direction that the rudder is being held and it is all over. What happened?

Let's climb to a safe altitude and duplicate the conditions that produced the spin.

1. The pilot flew a slow pattern. We'll slow the airplane to minimum speed and use whatever power is necessary to hold altitude. This means that the airplane is being flown at a relatively high angle of attack. Because of the induced drag, considerable power is necessary to hold altitude. Also, right rudder will be needed to correct for left yaw. (This is more difficult to see and react to in poor visibility.)
2. The pilot overshot the projected centerline and avoided a steep bank by using excessive rudder and opposite aileron. While still at low airspeed, we'll start a turn with excessive rudder. The use of this rudder will result in the nose dropping excessively. We'll correct for this tendency by pulling harder against the stick/wheel to get the nose back up. It is at this point that the angle of attack increases beyond that necessary for a given angle of bank. The use of opposite aileron to prevent the bank from steepening results in the depression of the aileron on the low wing. This means that the low wing has the greatest effective angle of attack. Also, because it is on the inside of the turn, it has the least airspeed.
3. Suddenly and without warning the airplane spun toward the rudder being held and the low wing. There was no warning by stall buffet, which is usually felt in normal stalls, because the low wing was essentially the only wing that was stalled. And probably the only portion of the wing that was completely stalled was the length of the deflected aileron. What is felt as stall buffet is the result of disturbed air from the stalled wing striking the tail surfaces. When only the outboard portion of one wing stalls, there is nothing for the disturbed air to buffet against.
4. Although most traffic patterns are flown to the left, and this hypothetical spin accident was probably the result of a left spin due to a skidding turn, another contributor to left accidental spins is left yaw due to the power

reaction at low speed. It is possible that our unfortunate pilot, while using considerable power at low airspeed during the turn, did not realize the extent of the left yaw due to power. During a turn, this yaw can be well disguised, especially under conditions of reduced visibility. It should be remembered that an airplane enters a spin quicker during high power settings.

THE IMPORTANCE OF SLOW-SPEED PROFICIENCY Being able to fly the airplane at a speed just above a stall is extremely important. The speed should be slow enough so that an occasional "nibble" into the stall takes place. When this occurs, it is quickly eliminated by a slight amount of forward stick/wheel with a minimum loss of altitude. Practice rolling in and out of as steep a turn as possible, using as much aileron deflection as possible. Of course, this experimentation should be done at a safe altitude.

Suppose you found yourself in a boxed canyon that you were unable to climb out of. Would you be able to make a quick, tight, steeply banked turn without entering a spin?

Skillful stick and rudder work during slow-speed flight and during the stall is the key to avoiding inadvertent spins. With practice, it is possible to fly most airplanes within a full stall and not only keep the wings level but roll in and out of turns without spinning. It takes practice, but it can be done.

A valuable coordination exercise that pilots should practice frequently is to roll the airplane right and left to various bank angles while preventing the nose from yawing. The nose may turn slightly into the turn during the first bank, but from then on it should remain centered upon some distant point.

The object is to develop proper yaw control as the nose tends to swing adversely, or opposite to the direction of bank. The amount of adverse yaw that will be experienced will vary with each airplane, but this maneuver will quickly reveal the yaw characteristics of a particular airplane. When flying a strange airplane, this is one of the first maneuvers I perform, usually during the climb out after takeoff, to become used to the control use necessary to properly coordinate the controls.

For maximum benefit, this maneuver should be practiced (at a safe altitude) at speeds slightly above the stall, where aileron drag is high and the adverse yaw most pronounced.

The pilot may experience an occasional stall at these speeds, usually as the angle of attack increases on the downgoing wing. Releasing a slight amount of back pressure against the stick at this time will quickly restore control of the airplane.

Stalls can also occur as a result of the downward deflected aileron increasing the effective angle of attack on that wing. The instructor should always point out the reasons for these incipient stalls as they occur.

For maximum benefit, this maneuver should be practiced from near-stall speeds up to the maximum allowable maneuvering speed of the airplane. This permits the pilot to experience the variable control pressures that develop during the maneuver throughout this speed range.

Pilots should guard against the tendency to become mechanical in their control use. This can be avoided if the maneuvers are practiced at various control rates, banks, and speeds.

THE RUDDER-EXERCISE STALL The exercise that most effectively develops a pilot's anti-spin instincts is the controlled stall, sometimes referred to as the "rudder-exercise stall." This maneuver is accomplished by using full up elevator when the stall occurs and holding it there. The ailerons are centered and only the rudder is used to maintain lateral control and prevent autorotation. If the airplane rolls to the right, left rudder is used to level the wings. If it rolls to the left, right rudder is used. With practice, you will be able to maintain reasonable lateral control and even execute turns while holding the stick/wheel fully aft. Quick stop-to-stop rudder activity will often be necessary to prevent a spin.

This is the kind of practice that develops both the perception of incipient spins and the reactions to avoid them. The distractions that result in stall/spin situations will not result in a loss of control for the pilot whose instincts automatically react to prevent autorotation.

During this type of practice, spin entries will occur until the pilot develops quick, automatic reactions against them. This in itself will develop better reactions and spin-recovery skills.

SNAP ROLLS The snap roll is an excellent practice maneuver to use to develop a feel for when the airplane is about to enter a spin from conditions of more than 1 *g*.

Snap rolls are nothing more than spins performed along the horizontal plane. As with any normal spin, rotation is about both the yaw and the roll axes. See Figure 5.4.

Because extra stress is imposed on the airframe, as well as on the engine, the engine mounts, and the propeller, only airplanes that are stressed and approved for aerobatics should be used.

Snap rolls should be performed at the entry speeds recommended by the manufacturer. The control application for the entry is essentially the same

FIGURE 5.4 Snap rolls: Spins along a horizontal plane.

as entering a spin from 1 *g*, except the controls are applied briskly and it is not necessary to pull the stick fully aft.

Although the controls are applied rapidly to produce autorotation at the higher entry speed, they should not be used roughly. Positive, firm application is all that is necessary. The rudder and stick/wheel should be applied simultaneously, but more rudder action is involved than stick action. The technique will vary somewhat from airplane to airplane, but yaw effort should slightly exceed pitch effort for brisk, clean snap rolls. Recoveries are also more positive because the maneuver is performed at a minimum angle of attack. The recovery is effected by use of rudder opposite to rotation. It may also be necessary to release a little back pressure from against the stick if excessive aft stick was used during the entry. Executed properly, very little, if any, back pressure will be released because the up elevator position will be the approximate up angle needed to hold the nose level with the horizon at the lower trim speed of the completed maneuver.

At these higher airspeeds, the controls produce more positive entries and recoveries, although rotational speeds will be faster.

The point during the roll at which recovery is initiated will vary with the particular airplane and its speed of rotation. However, recovery is usually started after the airplane rotates somewhat past the inverted position.

There are many variations of the snap roll, including half snaps, multiple snaps, and snap rolls performed in the vertical plane or as part of other maneuvers. The inverted snap roll is a variation of the inverted spin. However, in this book we will only be discussing the types of snap rolls that will be useful in developing the skills essential to preventing inadvertent spins. One of these maneuvers is the "vertical reversement."

THE VERTICAL REVERSEMENT

The vertical reversement is a partial snap roll that starts from a steep turn in one direction and concludes in a steep turn in the opposite direction. See Figure 5.5. This maneuver helps the pilot develop a feel for when a spin is about to occur while performing a steeply banked turn.

For practice purposes, enter a steep bank of about 60 to 70°. If it is not a well-coordinated, ball-centered turn, it will be difficult to perform maneuvers with consistent entry and recovery characteristics. The entry speed for the maneuver should be about 1.5 times the normal 1-g stall speed (this may vary with the airplane). The moment the airplane has a stable bank angle and airspeed, back pressure against the stick is rapidly applied simultaneously with top rudder (if you were in a left bank it would be right rudder). As soon as the airplane starts to snap over the top of the turn, opposite rudder is applied, along with a slight release of back pressure. If your timing is right, the maneuver will stop in a 60 to 70° banked turn in the opposite direction.

During early practice, you can expect to stop the maneuver in almost any attitude, ranging from upside down to straight down, or after you recover from a spin. With practice, you will be able to stop the maneuver exactly where you want to with precision and confidence.

If we always play a piano on a limited portion of its keyboard and practice only occasionally, we will never develop into accomplished piano players. The same is true in flying, although the only hazard that can prevail at the piano is possibly falling off the bench. Our lives, and the lives of those flying with us, depend upon our skills. If we fly sporadically and limit ourselves to a small segment of the total performance range of an airplane, our proficiency will become severely limited. We become proficient only in the things we practice, and we stay proficient only in the things we *continue* to practice.

If we never practice stalls and spins and never practice maneuvering an airplane near the stall speed, we eventually become incompetent in those aspects of flight. Then, if we are forced out of the routine aspects of flight, we can expect to be embarrassingly inept.

FIGURE 5.5 Vertical reversement.

It is a good idea to avoid stall- and spin-inducing situations at low altitudes. However, if you avoid practicing slow flight, stalls, and spins at a safe altitude, you may rob yourself of the automatic and instantaneous reactions that are essential to survive potential stall/spin situations at low altitudes.

Because of the nature of landing approaches, you are far more likely to fly inadvertently into the low-speed end of the airspeed range during an approach than you are to push the airspeed inadvertently to the red line during other flight situations.

Improficient pilots can easily be trapped into abnormal and dangerous flight situations at low altitudes by the following:

1. Allowing themselves to become distracted, e.g., by tuning radios or picking up dropped microphones
2. Looking for other aircraft
3. Maneuvering rapidly to avoid other aircraft
4. Aborting approaches
5. Misinterpreting ground speed for airspeed, especially during a tail wind or in high density altitude conditions
6. Flying into the area of reverse command during a slow approach, perhaps while trying to space yourself behind a slower aircraft
7. Adding power while in a low and slow steep turn and not correcting for power pitch-up, resulting in a stalling angle of attack
8. Using excessive rudder during the turn onto final approach, resulting in the need for opposite aileron and excessive up elevator, an especially deadly combination
9. Failing to correct for left yawing tendency during a steep climb after take-off or during an aborted approach

Pilots should indulge in enough flight training so that they do not have to think about maintaining control of their airplanes. Basic control should be so automatic and deeply ingrained that the mind is always free for the judgmental aspects of flying. Flying is indeed a thinking person's game, but no pilot is capable of exercising good judgment when constantly preoccupied with the manipulation of the controls and maintaining basic control of the airplane.

THE INADVERTENT FOGGY ROTATION (IFR) Forgetting to turn on the Pitot heat, the captain of a Boeing 727 climbed into a very turbulent overcast. It soon became difficult to prevent an increase in airspeed. Thinking there was a strong updraft, the pilot pulled the nose higher and higher to prevent the airspeed from exceeding the red line. Finally, much to the pilot's shock, the airplane entered a spin from which it was impossible to recover. The pilot had chased an erroneous airspeed indication all the way to a spin.

The Pitot head had iced over, trapping air within the airspeed instrument. The static port, free of ice, still sensed ambient pressure. As this pressure became less with altitude, the trapped air within the instrument expanded, recording an increase in airspeed.

There are a number of reasons why a pilot might enter an inadvertent spin while flying by instruments. For example, an airplane can become very difficult to fly with a load of ice and, because of a drastic change in mass distribution and aerodynamics, might present the pilot with a real problem in recovering from a spin.

Single-pilot IFR flying presents many possibilities for distraction that may result in the loss of control of an aircraft. Turbulence, darkness, numerous frequency changes, copying clearances, dropping pencils on the floor, trying to find the right approach plate, etc., are open invitations for inadvertent spins or other types of loss of control. A tired pilot can be as unresponsive as a mired mule to situations that demand quick and precise attention. Although I do not have statistics to prove it, I am convinced that pilot fatigue should be near the top of the list as a killer of pilots.

Those who have already had spin training under visual conditions will have a considerable advantage over those who have not. This is assuming, of course, that those who have had spin training have also had instrument training. Knowledge of the timing of recovery control use and the pressures felt both against the seat and against the controls will be of considerable value in learning to recover from a spin by reference to instruments. For those already trained in spins, the main task will be interpreting the instruments in relation to what is already known about spins.

The spiral is the most common out-of-control maneuver that is experienced when flying by instruments. The initial loss of control may have been instigated through some other maneuver, but the spiral seems to be the route to destruction that captures the inept, tired, or overtaxed instrument pilot.

Spins are not among the more common offenders in IFR flying accidents, but enough of them have occurred to make them a matter of concern. If the pilot is not distracted by other factors, the initial loss of control will first be revealed by the heading and attitude indicators. A rapid change in heading, combined with a rapid bank, may mean an entry into a spin. If the vertical-speed indicator reveals a rapid rate of descent, and the airspeed indicator shows a speed near or slightly over the stall, the airplane is spinning. If the airspeed is increasing rapidly, the airplane is in a spiral. In either case, the throttle should be closed.

Disregard the attitude indicator and observe the turn-and-slip indicator. The attitude indicator may not be reliable. For a spin recovery, push *full* opposite rudder to the direction of needle deflection. If the needle is deflected to the left, push right rudder. As soon as the rudder is deflected, move the stick/wheel quickly to the neutral position (or neutral elevator pressure). When the turn indicator centers and flicks momentarily opposite, the spin rotation has stopped. When this occurs, neutralize the rudder pressure. At this point the pilot needs to be concerned about a rapid increase in airspeed. The airplane will be pointed nearly straight down, and it takes only a few seconds before the speed is over the red line.

Apply firm back pressure against the stick/wheel while keeping the ball and needle of the turn-and-slip indicator centered by coordinated control use. Watch the airspeed. When the airspeed *starts* to decrease, the airplane is going through level flight. When this occurs, release the back pressure against the stick/wheel and permit only a slight climb rate on the vertical-speed indicator. As the airspeed drops toward the normal cruising range, smoothly apply climbing power while continuing to fly by the turn-and-slip indicator. (Watch for a left turn.) A minimum but positive climb rate should be established, and while climbing back to the original attitude, reset the turn indicator with the magnetic compass and watch for indications for the attitude indicator beginning to level. Establish a safe heading and hold it until the attitude indicator levels. This may take quite some time.

The recovery from a spiral should also be accomplished by using the same instruments. Disregard the attitude indicator. Except for stopping spin autorotation, the recoveries are identical.

SUMMARY

1. Miscoordination not only contributes to accidental spins but may result in spins that are difficult to recover from.
2. Skidding turns are among the strongest exciters of inadvertent spins.
3. All healthy turns are ball-centered and more difficult to spin from.
4. The rudder's primary purpose is to correct for the adverse yaw that results from unequal drag of the ailerons and for the asymmetric lift and drag that result from variations of angle of attack of the two wings. It should never be used by itself to increase or decrease the turn rate.
5. Downward aileron deflection can induce a stall at low speed as it increases the effective angle of attack on that wing.
6. Miscoordination can change the stall pattern of a wing and can result in a wing tip stalling before the root section. This frequently results in a lack of aerodynamic stall buffet and warning, as well as in a quick roll-off at the stall.
7. Normal spins incorporate both roll and yaw. When the rolling portion of a spin translates to yaw, a flat spin pattern develops.

8. In most cases, the use of aileron with or against the spin is not recommended.

9. In most cases, the use of power increases rotational speed, flattens the spin, and prolongs the spin recovery. However, a small amount of power is an aid to producing a positive and effective spin entry. However, the throttle should be closed as soon as the spin develops. A small amount of power is necessary in some airplanes to keep the engine running during a spin.

10. Most airplanes take from four to six turns to develop their full spin rate and dynamics. However, two turns will provide enough dynamics for practice purposes.

11. Always use the recovery techniques recommended in the pilot's handbook. This usually consists of rudder opposite to spin rotation, followed by forward stick/wheel. The rudder-prior-to-stick sequence is important in avoiding a speedup in rotation during the recovery, and a transition into an inverted spin.

12. Just as physical exercise improves physical health, proper flying exercises improve flying skills.

13. Inadvertent stalls and spins often occur after an engine failure, a result of improper maneuvering and poor coordination. Sometimes excessive maneuvering is the result of a pilot being unprepared for such an emergency. Simulated forced landing practice with a competent instructor at regular intervals is recommended. Pilots should be psychologically prepared to accept the fact that the airplane will be damaged during an off-airport landing during an engine failure. They should be prepared to immediately cooperate with gravity and maintain flying speed and control. The object is survival!

14. Snap rolls are closely related to spins. However, while spins impose very little stress upon the airplane's structure, snap rolls can impose considerable load. Only airplanes approved for aerobatics should be used for this maneuver.

15. Inadvertent spins can result from a lack of instrument flying proficiency, and from erroneous instrument indications due to an iced Pitot head. Also, flying with a load of ice can drastically deteriorate an airplane's performance and handling qualities.

16. During IFR conditions, the first indication of a spin will be a loss of heading control and a rapid loss of altitude. If the vertical-speed indicator reveals a rapid rate of descent and the airspeed indicator shows a speed slightly above the stall, the airplane is spinning. The same indications with an increase in airspeed reveal a spiral.

17. The turn-and-slip indicator is the only reliable instrument to use for a spin recovery. Rudder should be applied opposite to needle deflection. When the needle is centered, relax the rudder pressure and concentrate on recovering from the resulting dive. Apply firm back pressure against the stick/wheel and establish a slight climb rate on the vertical-speed indicator. Establish a safe heading and hold it until the attitude indicator levels.

THE HAIRY-GO-ROUNDS

FLAT SPINS

The forces that produce autorotation and the counterforces that resist it will determine the speed of rotation and the pattern of the spin.

The spin is initiated and continued through a considerable imbalance of lift between the two wings. An airplane with a favorable spin pattern and good recovery characteristics always has a greater amount of anti-spin forces than pro-spin forces acting upon it. This means that the only reason the airplane continues to spin is because the pilot holds the controls in a pro-spin position. If the pilot released the controls, the anti-spin forces would take over and the spin would eventually stop.

In most cases, the spin pattern and maximum rotational speed are developed after four to six turns of spin. The airplane rolls, yaws, and pitches about its center of mass. The actual center of rotation of a normal spin, however, is somewhat behind the nose, forward of the c.g., and off to one side.

When a spin becomes abnormal and starts to flatten, the nose rises from the normal attitude and the rolling portion of the rotation dissipates and converts into yaw. In other words, it is no longer rolling about the roll axis. Also, the airplane begins to rotate closer to the c.g. When the mode of spin is completely flat, the axis of rotation is very close to the c.g., although the rotation may appear eccentric. This could be the result of uneven autorotational forces caused by variations of lift on the advancing and retreating wing and of the autorotational forces contributed by the fuselage. In its spin research, NASA discovered that at very high angles of attack during a flat spin one airplane's fuselage was contributing more to autorotation than its wing. See Figure 6.1.

As centrifugal force develops during autorotation, the nose of the airplane will tend to rise. The amount that it will rise will depend upon the c.g., the

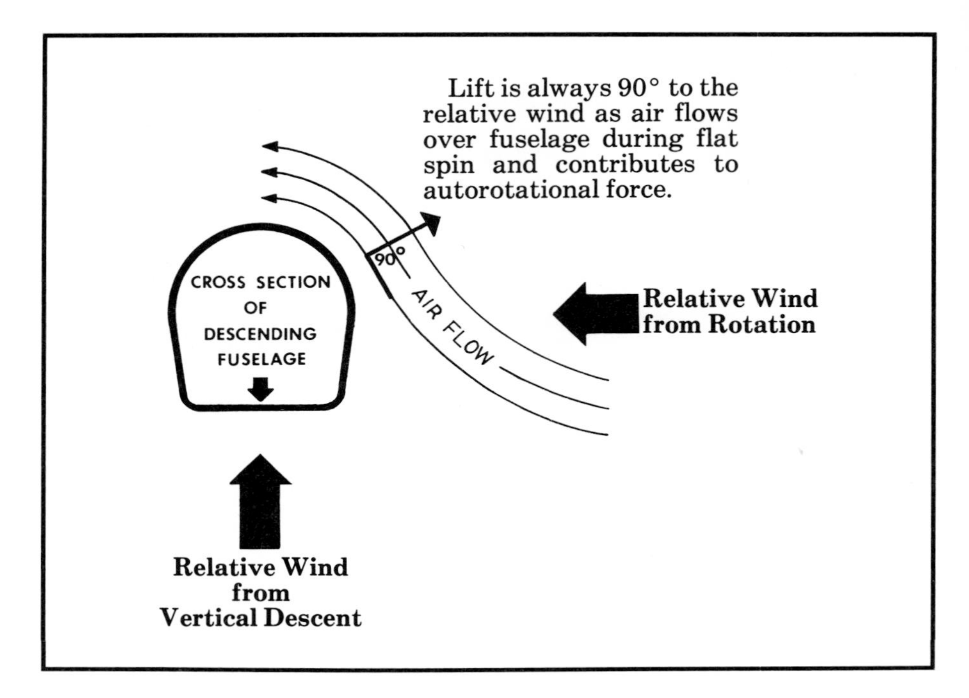

FIGURE 6.1 How the fuselage contributes to spin autorotation.

mass distribution, and the speed of rotation. The faster the rotational speed, the greater the tendency for the spin to flatten.

Although it is difficult to determine the angle of attack of a spin by observation, it has been my experience that it becomes difficult to recover from spins when the nose appears to be about 45° below the horizon. As the nose moves upward, recoveries become increasingly difficult. In its research, NASA arbitrarily identifies a moderately flat spin as occurring when the angle of attack reaches 50 to 52°. The flat spin mode is identified as occurring at an angle of attack of 70°.

As the spin flattens and the rotation comes closer to the c.g., there is less roll about the longitudinal axis and yaw becomes predominant. Under the influence of reverse flow, the retreating wing starts to develop lift and adds to autorotational force. By using aileron against the spin, the retreating wing, with the drooped aileron acting as a leading edge flap, provides additional lifting efficiency to the wing. The effect is now similar to a helicopter in autorotation. Both blades (wings) provide autorotational force. There is usually a dramatic increase in rotational speed when the retreating wing becomes a driving force.

During a flat spin, the vertical stabilizer becomes shielded from effective airflow. Also, its rapid sideward motion increases its angle of attack until it stalls and contributes little yaw-restoring effect. At very high pitch angles, the use of down elevator may further flatten the spin as it deflects all upward flow away from the vertical stabilizer. This is a technique used to flatten the spin of a ZLIN Akrobat, for example.

Many airplanes will not recover from a fully developed flat spin. However, if the spin is recoverable, it will be accomplished by closing the throttle, applying full rudder opposite rotation while using full aileron in the direction of the spin, and following this by full forward stick. The use of aileron in the direction of the spin spoils the lift developed by the retreating wing, and the downward-deflected aileron tends to provide drag against rotation and to recouple yaw into the spin. Jim Patton, a NASA research pilot, has had some success in breaking the pattern and recovering from some flat spins by pumping the stick fore and aft at the natural longitudinal frequency of the airplane.

It may take several turns for the airplane to recover. Patience and altitude are important. Leave the controls in the recovery position until the airplane responds. If it becomes necessary to bail out, leave the airplane on the *outside* of the spin (if the airplane is spinning to the right, leave on the left side).

It is interesting to note that although very few pilots have survived hitting the ground from a normal spin attitude, quite a few pilots over the years have survived ground contact from erect flat spins. The rate of descent is much slower during a fully developed flat spin.

The risk of being struck by the airplane while attempting a bailout under these conditions is extremely high unless the airplane is equipped with an ejection seat. Every emergency has its own set of circumstances and peculiarities, but it might be better not to attempt a bailout when an airplane reaches an extremely low altitude during a flat spin. It might be better to hold the airplane into the spin rather than attempt to recover from it. I'm not sure I'd follow my own advice under these circumstances, but it's worth tucking in the back of your mind for possible future reference.

As a rule, power increases rotational speeds and makes spin recoveries more difficult. Normally, the first step in a flat spin recovery is to close the throttle. However, if everything else fails, it may be worth a try. During a right spin, the precessional forces due to the propeller would tend to lower the nose (if they are strong enough to overcome other effects), and this could possibly be a method of breaking the stability of the spin and starting a recovery. I wouldn't count on the use of power being helpful however; it has been my experience that it only speeds up the rotation.

Although some airplanes will recover from flat spins, it should be remembered that a true flat spin develops the ultimate in pro-spin forces—with a *minimum* of recovery possibilities.

Flat spins, particularly inverted flat spins, are a common air-show exhibition maneuver. Some aerobatic airplanes display good flat spin recovery characteristics. Experienced aerobatic pilots, using aircraft of proven flat spin recovery characteristics, perform this maneuver routinely. Most of these aircraft must be held into this spin mode by the use of power; when the power is reduced, they tend to recover.

However, cautious exhibition pilots know that they are operating in a very critical area and take precautions that are important to longevity.

For example, the c.g. is kept within strict limits. Small aircraft, in particular, are extremely sensitive to small weight changes. A small change in pilot weight, for example, can account for a considerable shift in the c.g. Because the c.g. is calculated as a percentage of the mean aerodynamic chord (MAC), small wings and small chords mean a smaller allowable travel in the c.g. Quite often in small airplanes it is the pilot's weight alone that will determine whether or not the c.g. will be within safe limits.

There may be times during high density altitude conditions when pilots may want to exclude this maneuver from their routines because the higher rotational speeds and greater centrifugal loads may force the airplane out of its recovery range.

OSCILLATORY SPINS Some airplanes have a variable spin pattern resulting in the nose pitching up and down accompanied by variable rotational speeds. This often occurs as the result of entering a practice spin with a little too much airspeed. If oscillation occurs for this reason, it usually dissipates after two or three turns and settles down into a normal spin.

Some airplanes have an inherent oscillatory spin pattern. These are usually the result of the aerodynamic and gyroscopic forces not being in perfect balance. These spins usually rotate slower than the smooth, steady spin; as a result, recoveries are faster. Usually the best place to apply recovery controls is in the nose-low portion of the oscillation.

THE TRANSLATIONAL SPIN The normal spin has about an equal combination of yaw and roll. The flat spin is predominant in yaw. At an angle of attack of 90°, it is all yaw. During flat spins, rolling energy is converted into yawing energy; however, it is also possible to convert yaw into roll during a spin. This can be accomplished by pushing the stick forward while continuing to hold rudder in the direction of the spin. As the nose goes down and tail goes up, their arcs of rotation decrease in diameter. As a result of moving the rotating energy into a smaller circle (the way an ice skater does by pulling both arms inward when turning a pirouette), the rotational speed of the spin increases. If rudder is applied opposite to rotation in a steep nose-down attitude while this is taking place, the airplane will progress into an inverted spin while continuing to rotate in the same direction. It smoothly translates from an erect spin to an inverted spin. This is an important reason for using the rudder *first* during a spin recovery; it eliminates the possibility of translating into an inverted spin.

The experienced aerobatic pilot will recognize what is taking place and react properly. However, the inexperienced pilot may not recognize that a translation to an inverted spin is taking place. Pulling the stick aft under these conditions can convert the rolling energy into enough yaw energy to produce a flat spin. The recognition of whether the airplane is inverted or erect is the clue to a successful recovery. This can be difficult when the nose attitude is nearly vertical. However, once this is determined, rudder is applied opposite rotation to effect a recovery. The elevator is then applied relative to an erect- or inverted-spin recovery. If it is an inverted spin, the stick is pulled aft; if it is erect, it is pushed forward (because it is already forward, any further forward application may not be necessary).

OVER EASY

The Hollywood Hawks air show troupe had been pushing a cold front eastward all the way from California. They arrived at the old Curtiss-Wright airport on Long Island 20 minutes late, and the scheduled air show was awaiting their arrival.

The promoter, who had been dealing with an impatient crowd, asked me to get an act into the air as soon as possible. I frantically unloaded the baggage from my 450-hp Pratt & Whitney Stearman biplane, hooked up the electrical connections to a rocket-assisted takeoff unit mounted on the belly of the airplane, fired up the engine, and rapidly taxied into position for takeoff.

The act I was about to perform consisted of firing the rocket as soon as I had rudder control during the takeoff roll, then jerking the airplane into a vertical climb that continued to about 1200 ft. It was noisy, smoky, and spectacular. The big Pratt & Whitney, boosted by 1000 lb of rocket thrust, provided plenty of power for the maneuver.

As I hurriedly shoved the throttle forward, I forgot to switch the fuel valve to the all-attitude fuel system. After about 100 ft of takeoff run, I punched a button on the stick. The rocket fired and shoved me back into the seat. I pulled firmly against the stick, and the nose of the Stearman swept upward. The radial cylinders were etched clearly against the blue sky. I turned my head to the left and held the biplane's wing tips perpendicular to the ground.

I had to hold the stick well forward to compensate for the tendency of the airplane to arc toward inverted flight into a loop. The low thrust line of the powerful rocket tended to push the nose upward and around. Suddenly the faithful Pratt & Whitney quit cold! The airplane arced upside down. I jammed the stick full forward in an attempt to stop it. There was a moment of wallowing as the Stearman settled backward into the dense rocket smoke. Then the machine twisted, my weight went against the belt, and the machine spun inverted . . . nose high!

Smoke filled the cockpit and completely obscured my vision. I was spinning upside down through the dense column of smoke! I pushed hard rudder against the spin and pulled the stick into my stomach. The spin stopped; I broke free of the smoke and faced the frighteningly clear features of small trees, grass, and a taxi strip. Suddenly, the engine awoke from its untimely sleep, and, facing its terrestrial enemy, it roared vengefully. The combined rocket and engine power provided the energy and speed that I needed for control. Some low trees flashed by the right wing as I skimmed the grass with the wheels.

The show had started spectacularly; everything that followed would be anticlimactic. I received a standing ovation as I taxied up to the grandstand. Personally, I didn't feel like standing. I sat quietly in the seat for a few moments, gathering enough strength to climb from the cockpit. Witnesses said that I couldn't have been higher than 300 ft at any time and that the recovery arc, considering my proximity to the ground, looked unreal.

If people had told me that I would someday find myself in an accidental inverted spin at 300 ft, I would have said they had cracked their marbles. Yet that was my surprise package about 32 years ago. It was a well-learned lesson that haste and carelessness breed low-altitude spins, and sometimes waste.

Unless you're an air show performer, or are flying beyond your proficiency at low altitude, you'll probably never get into an accidental low, inverted spin. But it isn't always the foolhardy person who gets into trouble; quite often it's the cautious Charlie who never exceeds a 20° bank at any altitude.

AN ALL-ATTITUDE VEHICLE Because an airplane is an all-attitude vehicle and can be flown in any attitude, inadvertently or on purpose, aerobatics should be part of every pilot's training—and that includes both erect and inverted spins.

I realize that this view is highly controversial, but I am convinced that aerobatic training not only improves pilots' flying abilities and increases their confidence, but also makes them safer pilots. We wouldn't need all of the safety campaigns and regulatory Band-Aids if pilots were properly trained in the first place.

During an inverted spin, the wing, the motivator of all spins, is upside down, and when one wing becomes stalled because of an excessive inverted angle of attack, inverted autorotation takes place. The pilot, of course, hangs against the seat belt and shoulder harness. See Figure 6.2.

FIGURE 6.2 Pilot's perspective of an inverted spin.

INVERTED FUEL SYSTEM Unless the airplane is equipped with an inverted fuel system, the engine will quit soon after the airplane reaches the inverted position. The propeller will continue to windmill as long as airspeed is maintained, but when the airplane is slowed to reach a high angle of attack, the propeller may stop turning and will need to be restarted after recovery from the spin and return to erect flight. It is far better to select an airplane with a full inverted system for this kind of flying.

THE AEROBATIC HARNESS Some seat belts would fail while holding your pants up, let alone holding you firmly in the seat while upside down. I discovered this the hard way while attempting an inverted spin in an old "ragwing" Luscombe that was fitted with a "sure slip" belt.

I rolled inverted, shoved the nose up, and pushed the controls into an inverted spin position. Just as the airplane slung itself into an inverted pirouette, and the seat belt began cutting into my stomach, I suddenly left the seat and my head went crashing through the overhead skylight. I still hung onto the stick, pulling it aft as my feet left the rudder pedals. The spin stopped, and as positive *g*'s developed, I fell back down into the seat. Fortunately, the skylight wasn't large enough for anything larger than my head to get through; otherwise I would have hurtled toward the ground without an airplane, a parachute, or a future.

As I maneuvered the airplane back into level and sane flight, I sorted the glass out of my hair and concluded that a full aerobatic harness was an absolute necessity if I was going to continue experimenting with inverted maneuvers.

An aerobatic harness should consist of two seat belts and a good shoulder harness. The seat belts should be reasonably tight and the harness slightly more than snug. Most of the pilot's weight should be borne by the seat belts when flying inverted. If the seat belts are not tight enough and all the weight is being borne by the shoulder harness, it may place unnecessary stress upon the spine. However, the shoulder harness should be tight enough to prevent excessive stretching of the upper torso. This stretching can make it difficult to reach the stick while being subjected to negative *g*'s.

It has been my experience that some inverted spins appear flatter and seem to rotate faster than erect spins. Sometimes these negative loads make it more difficult to hang on to the stick.

Flat inverted spins also tend to create a reversal of elevator stick forces, so that the stick tends to pull forward out of the pilot's grasp. There is nothing quite as embarrassing as having the stick full forward and being unable to reach it to effect a recovery. Be sure that your seat belt and harness are tight enough!

Tex Rankin used to tell the story about losing his grip on the stick while performing a flat inverted spin at an air show. He removed his feet from the rudder pedals and attempted to corral the stick with his feet. He was finally able to pull the stick far enough aft with his legs and feet to again grasp the stick with his hand. The Great Lakes biplane was now so low to the ground that he had to recover from the resulting inverted dive by "pushing under" and climbing inverted until he was high enough to roll upright.

Now and then even experienced air show performers spin all the way to the ground while performing inverted flat spins. I wouldn't be surprised if some of these accidents were due to the pilots simply losing their grip on the stick. Stick reversal (the tendency for the stick to remain forward without being held there) is common during inverted flat spins; it may take a strong force to pull the stick aft.

For a given rotational speed and nose attitude, some airplanes seem to recover easier from inverted flat spins than from erect flat spins. This is usually attributed to less aerodynamic shielding of the vertical fin and rudder while inverted (see Chapter 3, Figure 3.9). However, the autorotational forces contributed by both wings and the fuselage may overpower the unshielding effect and recovery may become impossible.

Like erect spins, inverted spins can be entered from any attitude. However, normal practice entries are usually from inverted flight. After rolling inverted, the power is reduced to idle and the stick is gradually pushed forward until the first nibble of stall is felt. At this time full rudder is applied in the direction of the intended spin and the stick is pushed well forward. In most airplanes a more positive entry can be obtained by using a little opposite aileron at the time of rudder application. (This is aileron *with,* not against the spin.) As the spin develops, the aileron is again centered.

The controls are held in this position until a recovery is desired, at which time the rudder is pushed briskly opposite spin rotation, immediately followed by pulling the stick quickly aft. As rotation stops, the rudder is neutralized and the airplane is brought back to level flight by pulling up from the resultant dive. The nose will swing toward a positive-*g* arc as the stick is pulled aft during the recovery.

CORIOLIS EFFECT

Using opposite rudder and aft stick simultaneously will effectively stop an inverted spin in most airplanes. However, in some airplanes this technique results in a "Coriolis effect," which is a result of reducing the spin radius as the stick is pulled aft. This couples yaw forces into roll forces and delays the recovery. The sequence should be the same as for recovering from an erect spin and for the same reasons. Use rudder first, followed by stick/wheel.

Remember that holding the controls in the anti-spin position until the rotation stops is extremely important. In some airplanes, and under certain conditions, this may take several turns.

Years ago, while conducting spin tests on a jet trainer, I encountered a severe inertia-coupling effect as I simultaneously used rudder and stick to stop an inverted spin. The speed of rotation increased so dramatically that even though my seat belt and shoulder harness were tight before I entered the spin, I was thrown away from the seat and my head was pushed sideways against the canopy, my face resting against the top of the canopy.

I did not realize it at the time, but as the spin rate increased, I pushed the rudder back into the direction of rotation, further increasing the rotational speed. This was later revealed by the oscillograph records. Knowing I was in trouble, I attempted to reach the spin chute deployment lever, which was located to the left of my seat, on the floor. It was impossible; I was being thrown away from it by centrifugal force.

The dysphoria was so overwhelming that I was unable to function physically or mentally. I said a brief prayer and waited, perhaps relaxing my efforts against the controls. After several more turns, the rotational speed decreased, the nose lowered, and I was able to take over and further the recovery.

If you become confused about the spin direction, look at the turn indicator and apply rudder opposite to the needle deflection. The ball portion of this instrument means nothing during a spin.

Although the above makes a good war story, there is really very little to be concerned about in an approved aerobatic trainer with a competent instructor. An introduction to inverted flight is essential before attempting inverted spins. You will need to become orientated in this type of flight before the full value of inverted spin practice is realized.

SUMMARY

1. Flat spins result when the inertial forces overpower the aerodynamic pitch-down forces.
2. Yaw predominates over roll during a flat spin. At a spin angle of attack of 90°, there is no roll.
3. As a rule, it is difficult to recover from spins when the nose is within 45° of the horizon.
4. NASA arbitrarily identifies a moderately flat spin as occurring when the spin angle of attack reaches 50 to 52°.
5. As the spin flattens, the rotation becomes closer to the c.g.
6. Many airplanes will not recover from a fully developed flat spin.

7. The recommended recovery procedure is to close the throttle, apply full rudder opposite rotation while using full aileron in the direction of the spin, and follow by pushing the stick full forward.
8. Flat spins develop the ultimate in pro-spin forces with a minimum of recovery possibilities.
9. Flat-spin prevention starts with keeping the c.g. within limits.
10. Small aircraft, in particular, are sensitive to small weight changes. Often, it is the pilot's weight alone that will determine whether or not the c.g. is within limits.
11. Oscillatory spin patterns are caused by an imbalance in aerodynamic and gyroscopic forces.
12. Translating from an erect to an inverted spin becomes possible when forward stick is applied before recovery rudder.
13. A good shoulder harness and double seat belts are important equipment that should be installed before attempting inverted spins.
14. Use airplanes equipped with inverted fuel-and-oil systems for inverted spins.
15. The normal recovery from an inverted spin is accomplished by using rudder against rotation followed by pulling the stick aft.

7

"*A fool and his money are soon flying more airplane than he can handle.*"

-P. T. Barnstorm

DOUBLE TROUBLE

TWO ENGINES, TWO NATURES The dual-engine airplane also has a dual nature. With two healthy engines, it performs and responds like a stallion. However, when one engine develops palpitations of the pushrods, it can be as deviate and distressing as a decapitated duck.

After an initial exposure to the single-engine exercises required for a multiengine rating, most pilots of twin-engine aircraft concentrate only on the multiengine mode of performance and handling characteristics.

The false cult of twin-engine reliability has led many pilots astray. It is true that the modern aircraft engine is reliable, and you can expect two of them to run at the same time most of the time. However, if one of them fails, you've lost more than half of the airplane's performance; in fact, the performance will be reduced by about 80 percent. Under these circumstances, the remaining engine is often overworked and overheated and the possibility of its failure is increased considerably.

When experiencing engine failure, the pilot of a single-engine airplane knows that it is necessary to cooperate with gravity and glide to the best possible landing site. However, if the pilot of a twin experiences failure of one engine soon after takeoff, that pilot may have entered the indecisive realm of hoping for the best. It is here that fatalities are bred. The single-engine margin of performance of the average light twin is inadequate to accept anything but the most exacting flying skills. All too often the person at the controls is proved deficient in these skills.

Pilots of single-engine aircraft are far more likely to be aware of the necessity of combining skill and judgment with the threatening forces of drag and gravity than are those who have spent many hours with the anesthetic drone of two engines. Somehow, the extra engine gives pilots the feeling that they have extra protection deposited in the bank of safety. However, unless they take the time to practice in the single-engine realm, they may find their accounts quickly overdrawn at a time when they need safety most.

SUDDEN-SPIN SYMPTOMS It has been proved that an engine-failure-related accident is twice as likely to produce a fatality in a twin than in a single-engine airplane. One of the reasons for this is that the loss of an engine can suddenly produce all of the situations that we have been discussing in regard to inadvertent stalls and spins. Pilots are suddenly exposed to asymmetric lift, variable angles of attack between the two wings, a very high amount of induced drag, and, often, a lethal amount of yaw. Because most of their flying has been under the influence of better than adequate performance, they lack the skill and judgment to cope with a very dangerous situation.

Of all pilots, the pilots of twins should certainly have spin training. Unfortunately, all twins are restricted against spins, some of them for a very good reason: their spin characteristics are dangerous. However, nearly all airplanes can be prevented from entering spins if the conditions that result in autorotation are recognized soon enough. For this reason, I strongly recommend that pilots of twin-engine airplanes receive spin training in an airplane approved for spins. I cannot think of anything that will bamboozle pilots' aeronautical audacity quicker and more completely than experiencing their first spin, unwillingly, in a double-engined perplexer.

Before discussing how to recover from a spin in a twin, let's consider the situations that lead up to autorotation in these airplanes.

First, let's look at how the power is distributed. Each engine is mounted just outboard of the fuselage. The slipstream from the propeller flowing backward over the wing contributes significantly to lift. The pilot must keep power and lift in balance by using the throttles. A significant variation of power between the two engines will result in unwanted roll or yaw. How the power is distributed will affect the controllability of the airplane. The two throttles are actually part of the control system of a twin. For example, twin-engine pilots frequently use asymmetric power to control yaw during a crosswind landing. They lower a wing to stop the drift and apply asymmetric power to keep the airplane aligned with the runway.

Can you imagine what would happen in a single-engine airplane if the pilot applied full rudder when the engine failed during a climb after takeoff? Sounds ridiculous, doesn't it? Because power also controls yaw in a twin, this is essentially what can happen the moment one engine quits at slow speed on a twin-engine airplane. Unless the pilot performs some quick and precise rudder work, a dangerous amount of yaw will occur. In addition, half the power loss will result in a loss of approximately 80 percent of the airplane's performance—this, at a time when airspeed is needed for both control and climb performance. At this point, the pilot must develop self-discipline in maintaining control of the airplane, because a loss of airspeed will result in a deterioration of yaw control. The natural tendency is to jam the throttle on the operating engine all the way forward, resulting in less rudder effectiveness in controlling yaw.

THE ESSENTIAL SACRIFICE The most difficult thing for a pilot to do when the airplane has lost most of its performance after takeoff is to sacrifice power for controllability. Yet, this must be done when a fully deflected rudder (and recommended bank) can no longer control the yaw. As in any emergency situation, the goal is survival. In order to maintain control of the airplane, there are times when the power on the operating engine *must* be reduced.

Pilots should know before taking off whether or not the airplane will have enough performance to continue climbing if it should lose one engine. They should know under what conditions of weight, density altitude, and power available the airplane will no longer climb. Then they will *know* that they must execute an off-airport emergency landing if one engine quits. The only possible use of the good engine would be to stretch the approach to a selected off-airport landing area.

MENTAL CONDITIONING The pilots' mental attitude during an emergency is extremely important. They should convince themselves beforehand that damaging an airplane during an off-airport emergency landing is not a reflection upon their flying ability. More than likely, damage will occur. However, if they maintain control up to the moment of ground contact, the chances for the aircraft's occupants surviving have been greatly increased. If they lose control of the airplane and spin in, survival is next to impossible.

V_{MC} IS NO HAVEN The minimum airspeed at which a pilot may control yaw with one engine dead and the other developing maximum power is known as V_{MC}. This speed is determined by factory test pilots with the critical engine (usually the left engine) inoperative.

It is usually specified that the minimum directional control speed be no greater than 1.2 times the normal 1-g stall speed in the lightest takeoff configuration. If an attempt is made to fly the airplane with a full imbalance of power below the minimum control speed, an uncontrollable yaw into the dead engine is assured. The only way control can be regained is to reduce power or sacrifice altitude for the sake of regaining airspeed.

Pilots need to be reminded that V_{MC} is not an area of haven, but a realm where disasters often occur. It's a domain explored by experienced test pilots under the most ideal conditions. The manufacturers, intent upon selling airplanes, have seen to it that their airplanes' V_{MC} reflects the lowest possible airspeed.

It is one thing for an experienced test pilot to probe the V_{MC} realm under controlled conditions, and quite another thing for an inexperienced pilot to experience an engine failure unexpectedly during an IFR departure at night or under other adverse conditions.

A lack of awareness of the hazards involved in the V_{MC} realm has resulted in an improper approach to twin-engine flight training. Approximately one-third of all accidents in light twins can be attributed to flight-training accidents.

Even when an airplane is flying well above minimum control speed, the sudden yaw and roll and resultant large control deflections caused by the suddenly asymmetric power can induce large amounts of drag and cause deterioration of airspeed to below the V_{MC} level. Also, power surges often precede the death of an engine. This results in momentary propeller overspeeds that produce large amounts of drag and further contribute yawing into the failing engine.

Inadvertent spins can occur during the V_{MC} phase of twin-engine operation. In normally aspirated engines, horsepower diminishes with increased altitude. This results in a weakened yaw reaction from asymmetric power. This means that the yaw can be controlled at a lower indicated airspeed. As altitude increases, the minimum control speed comes closer to the stall speed until, finally, the two meet. A spin is almost a certainty when a pilot loses directional control while entering a stall. Also, at these higher altitudes, the forces that induce autorotation produce a more violent spin entry in the less dense atmosphere.

An extremely aft c.g. can increase the V_{MC} speed. The distance between the c.g. and the rudder is reduced, decreasing the effective moment arm and reducing its effectiveness. Spins are more easily excited at aft c.g. because it is easier to obtain a stalling angle of attack. Also, as a rule, flat spins are more easily induced at aft c.g.

We have already considered how the use of ailerons without adequate yaw control can induce a spin. It is common (and even recommended practice) to use up to 5° of bank toward the good engine to reduce the minimum control speed. Although this is essential to controllability and rudder effectiveness in correcting for yaw, it can become extremely hazardous if overdone. If a pilot senses a tendency to hold full rudder and to use increasing amounts of opposite aileron, it is time to reduce power on the good engine. If power is not reduced, the pilot is in for a quick and unannounced spin entry. The depressed aileron creates an effective increase in angle of attack on a wing that is already shy of lift because of the inoperative engine. The other wing, under the influence of considerably more airflow from the propeller of the operating engine, is producing a great deal more lift. When the wing containing the inoperative engine finally stalls, the effect of considerable asymmetric lift results in a rapid rolling motion toward the dead engine.

One of the important disciplines that the pilot of a twin should develop is to be ready to reduce power on the good engine or sacrifice altitude in favor of airspeed, or do both, when nearing the limits of rudder effectiveness.

PRO-SPIN MASS DISTRIBUTION To understand the cantankerous spin nature of the average light twin and why they have a reputation for difficult spin recoveries, we need to consider the configurations that contribute to the problem.

Because most of the weight of a twin is distributed along its wing in the form of structure, engines, and fuel, it will induce certain gyroscopic properties that are strongly pro-spin. See Figure 7.1.

For example, as autorotational forces develop, the predominant force is that which develops about the roll axis, because the wing is the heaviest part of the airplane. At least during the incipient phase, the airplane rolls about the roll axis, yaws about the yaw axis, and pitches about the pitch axis.

When the dynamics of a strong rolling moment are coupled with pitch, a strong yawing moment is induced. This moment can become strong enough to couple the rolling energy into pure yawing energy, in which case a flat spin develops. An early recovery reaction is imperative in a twin, and this is only developed through actual spin recovery practice in an airplane approved for spins. If a pilot must think through the recovery procedure before taking action, it may be too late.

SPIN-RECOVERY POSSIBILITIES However, for those who refuse to take spin training, I include the following information with the hope that it will save their lives and the lives of their passengers. Let's hope they will be high enough to fumble through a recovery!

1. The instant the airplane rolls and yaws opposite to rudder and aileron, pull both throttles closed.

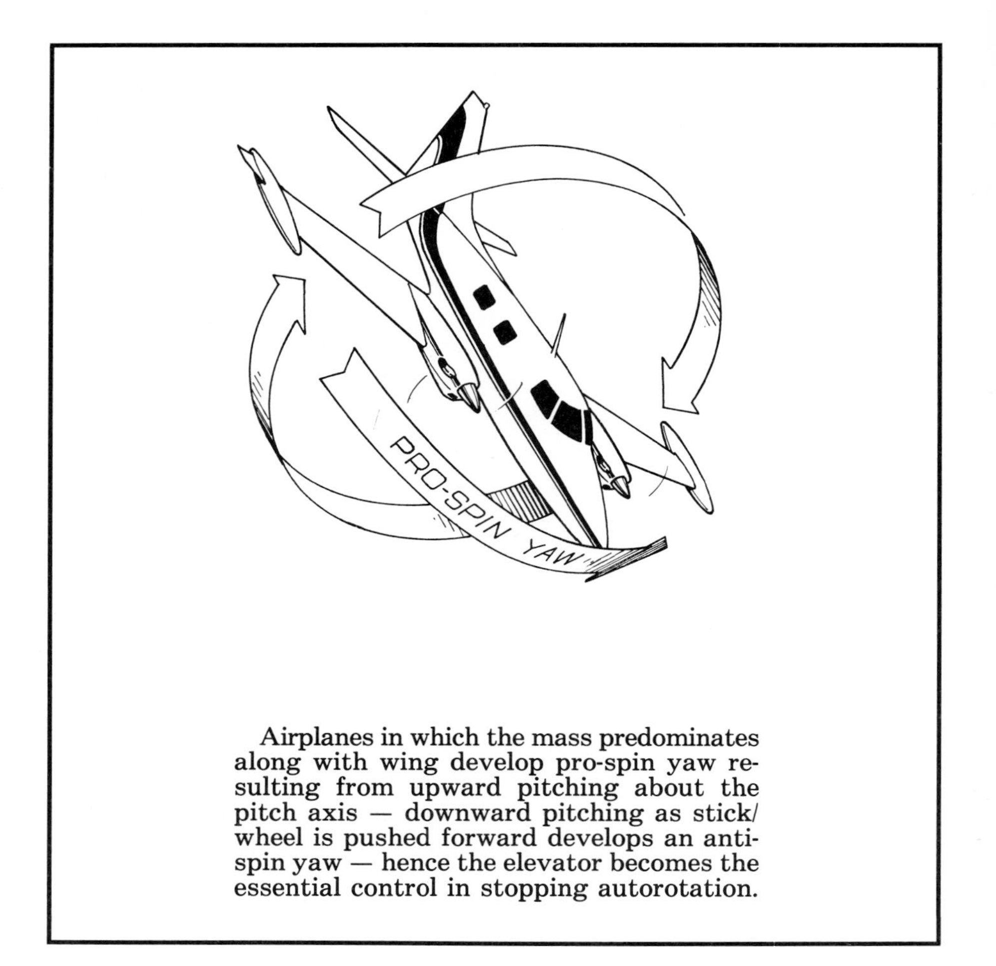

FIGURE 7.1 Gyroscopic effect due to predominant mass along the wings.

2. Center the ailerons and briskly apply full rudder *against* the spin rotation. During the incipient phase, the direction of rotation should not be difficult to determine. If the rudder is ineffective in stopping yaw, briskly shove the wheel all the way forward. If the rudder is responsive, full forward wheel will not be necessary. Continue to hold the controls in this position until rotation stops.

3. If the airplane does not respond to the controls and appears to be spinning flat (within 45° of the horizon), apply full aileron in the direction of the spin. This action will tend to spoil the lift on the retreating wing and reduce its autorotational contribution. This is a recommended course of recovery action in a flat spin in any airplane. See Figure 5.2*b*.

If power is available (engines tend to quit during flat spins), it may be worth trying asymmetric power against the direction of rotation (left spin, left throttle), although during a fully developed flat spin it may only increase the rotational speed.

If the airplane is spinning with the nose near the horizon in an obviously fully developed flat spin, and if the ground seems to be arriving fast and there is little hope of a dive recovery even if you could stop the spin, it would be well to maintain as flat a spin as possible. This is accomplished by using aileron opposite to rotation and adding full power. The rotational speed may increase, but it will assure a minimum descent rate. People have been known to survive flat-spin accidents.

Although I don't recommend experimentation in this area, except under controlled test conditions, I have experimented with recoveries from early incipient spin phases in light twins and found no difficulty effecting a recovery. However, those untrained in spins may not react quickly enough to prevent a full-blown, unrecoverable spin. The secret of spin prevention is enough actual spin practice to develop quick reactions against inadvertent spins.

SUMMARY

1. Pilots tend to acclimate themselves only to the twin-engine mode of performance and handling characteristics. As a result, they are totally unprepared when one engine fails.
2. The performance can deteriorate as much as 80 percent of its twin-engine capabilities.
3. Statistics show that an engine-failure-related accident in a twin is twice as likely to produce a fatality as one in a single-engine airplane.
4. The loss of an engine can suddenly produce all the conditions that can excite a spin.
5. The discipline of sacrificing the remaining single-engine power for controllability is often the key to survival. When the airspeed drops to near or below V_{MC}, the pilot should be prepared to reduce power to maintain control, and make an off-airport landing.
6. Pilots should *know* before they take off whether or not the airplane is capable of climbing under conditions of weight and available power, should it lose an engine. Attempting to remain airborne after an engine failure could be fatal.
7. Pilots should know that damaging an airplane during an emergency, off-airport landing is probable and professionally acceptable, and is not likely to be avoided. The object is survival. Attempting to remain air-

borne under questionable circumstances may make that impossible. The FAA should never criticize the pilot who elects to land rather than continue on one engine (this has happened).

8. Pilots should be reminded that V_{MC} is not an area of haven, but a realm in which disasters occur.
9. A spin is almost a certainty if a pilot loses directional control when entering a stall.
10. Aft c.g.'s can increase the V_{MC} speed.
11. Flat, uncontrollable spins are most easily induced at aft c.g.
12. Be wary when using aileron into the good engine to maintain directional control. The use of aileron can more readily excite a spin when control is lost.
13. Because of a pro-spin gyroscopic reaction to pitch-up, the heavy mass loading of a twin's wing subjects it to flat spins. Down elevator is the predominant control during a spin recovery, although for other reasons the rudder should be applied first against rotation.

"If we are what we eat, then some pilots should eat more chicken."
-P. T. Barnstorm

8 PROBING PARTICULAR PROBLEMS

Occasionally I have been asked how to investigate the spin characteristics of new airplanes. This section is written for those brave souls who have built their own airplanes and want to know how they will spin and recover.

If a comprehensive spin investigation has already been completed on a specific airplane design and the results were favorable, all that is necessary is a simple program to determine how a specific airplane compares to the prototype and previously determined standards.

TESTING DISCIPLINES However, spin testing of prototype models, if done properly, involves a methodical, step-by-step program. Safety demands that the test proceed slowly and that time be taken between each flight to analyze the results. Those who have done numerous spins in proven aircraft may have a tendency to approach the spinning of a prototype with a false sense of security. Spin testing is one of the most hazardous areas of flight testing, and pilots can unexpectedly find themselves in serious trouble if they do not approach their tasks with the utmost caution. Good test pilots work at their own pace and do not permit others to hurry them. There are always fools ready to risk their lives seeking selfish glory. The problem with these individuals is that they put pressure on more cautious people to act against their best judgment. I have seen pilots do foolish things despite their usual good judgment because there were other pilots ready to take their jobs if they didn't take the risk.

Sometimes it takes more courage to refuse a risk than to accept it. It certainly takes more integrity. There is always a safer way to do anything, and it is worthwhile to approach test flying with extreme caution.

THE CHASE PILOT Prototype spin testing should not be conducted without the aid of a skilled chase pilot. The pilot who will be performing the spin tests should select someone trusted to fly within reasonable proximity to the airplane being tested and observe the spin characteristics. Obviously, the chase plane should have adequate performance and maneuverability. Good two-way radio communication is absolutely essential. The chase pilot will also watch for other traffic and warn the test pilot when the test airplane is getting too low. The chase pilot should know enough about spins to recognize one that is potentially unrecoverable and should be able to instruct the test pilot in recovery procedures or spin-chute deployment operations should the test pilot appear unable to perform. The chase pilot should also be knowledgeable in ejection or bailout procedures and be able to relay this information to the pilot, should the need arise.

PHOTOGRAPHIC COVERAGE A camera operator should be part of the chase crew to take movies of the spin tests. These can be invaluable in analyzing the spin patterns.

AUXILIARY SPIN RECOVERY DEVICES Let me say at the outset that no airplane should be spin-tested without first installing an auxiliary anti-spin device that will stop the spin if conventional controls become inadequate. There are two such devices: wing-tip rockets and conventional anti-spin parachutes.

NASA has been experimenting with hydrogen peroxide wing-tip rockets that control roll and yaw. They are used to change the roll and yaw patterns of the spin and have been used more than the spin chute to stop unrecoverable spins. The spin chute is used more as a backup recovery device. It is mounted well aft of the tail in a tubular container. The deployment mechanism is a slug gun and a pilot chute that extract the spin chute clear of the wake of the spinning airplane. At the time of this writing, the device had been used 17 times to stop flat spins.

Most of us do not have access to the type of anti-spin equipment that is available to NASA. Nevertheless, simpler installations can be effective in stopping unrecoverable spins. However, because your life is at stake, you should use the best technology and equipment available to you.

I have used spin chutes on four occasions to stop unrecoverable flat spins. They were relatively simple installations; they were deployed by opening the pack container, which permitted the pilot chute to drag the main canopy, shroud lines, and drag cable to their full length.

Figure 8.1 shows a simple spin-chute installation which, in several variations, has been used to stop spins on many airplanes. This arrangement has worked well for me on three airplanes; one of the three was the T-33 jet trainer. I used the same basic configuration on several other airplanes that I tested, but never had to deploy them because the spins were recoverable without the spin chutes.

It should be understood that the spin-chute package itself can change the spin characteristics. The aft weight, even though offset by forward balance, can produce inertial effects that can be detrimental to the character of the spin. The structure should be kept as light as possible, but still strong enough to withstand the opening and drag loads of the parachute. Also, the aerodynamic profile should be kept as low as possible. The shape of the pack can improve or destroy the spin characteristics. If the spin-recovery characteristics are critical, the pilot should be wary when attempting to spin the airplane with the spin chute removed. See Figures 8.2 to 8.5.

A length of steel cable long enough to clear the tail should be installed between the nylon drag cable and the locking hook. The reason for this is

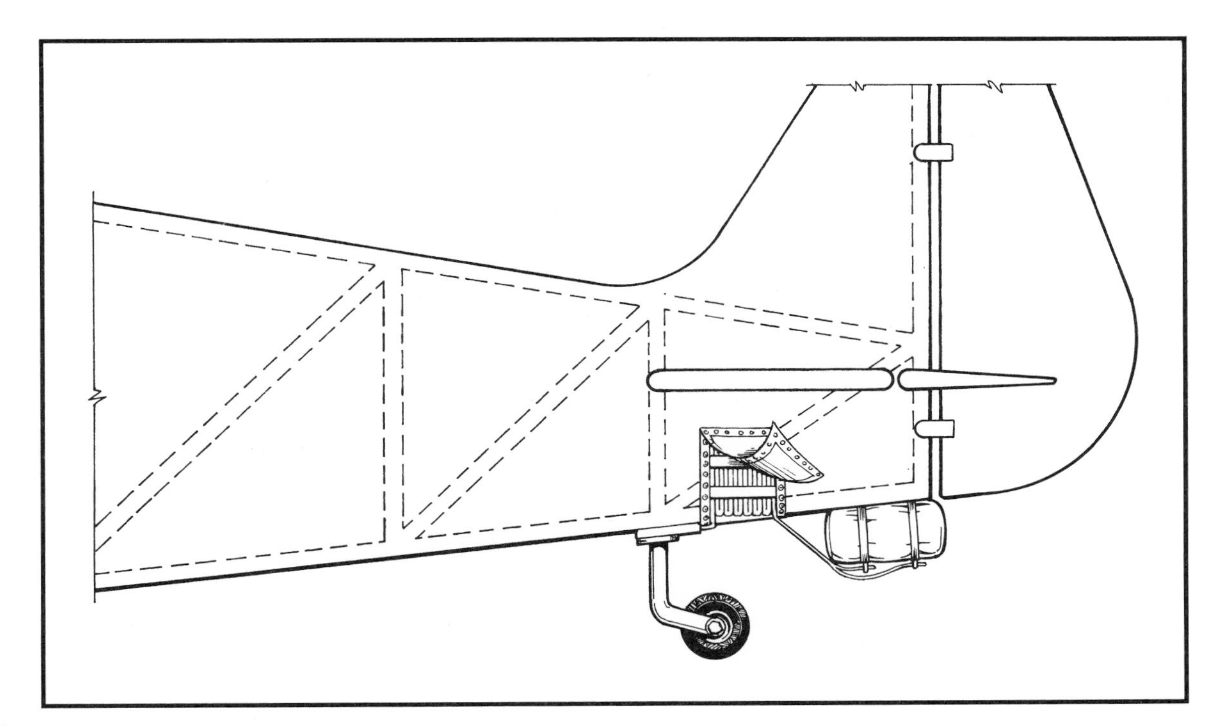

FIGURE 8.1 A simple spin-chute installation.

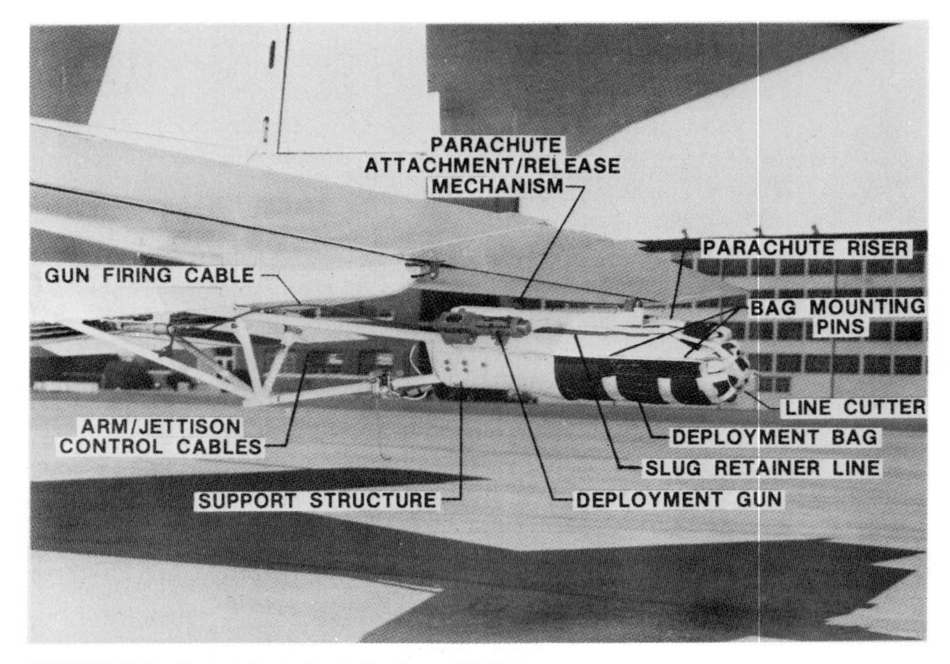

FIGURE 8.2 Spin-chute installation. *(NASA.)*

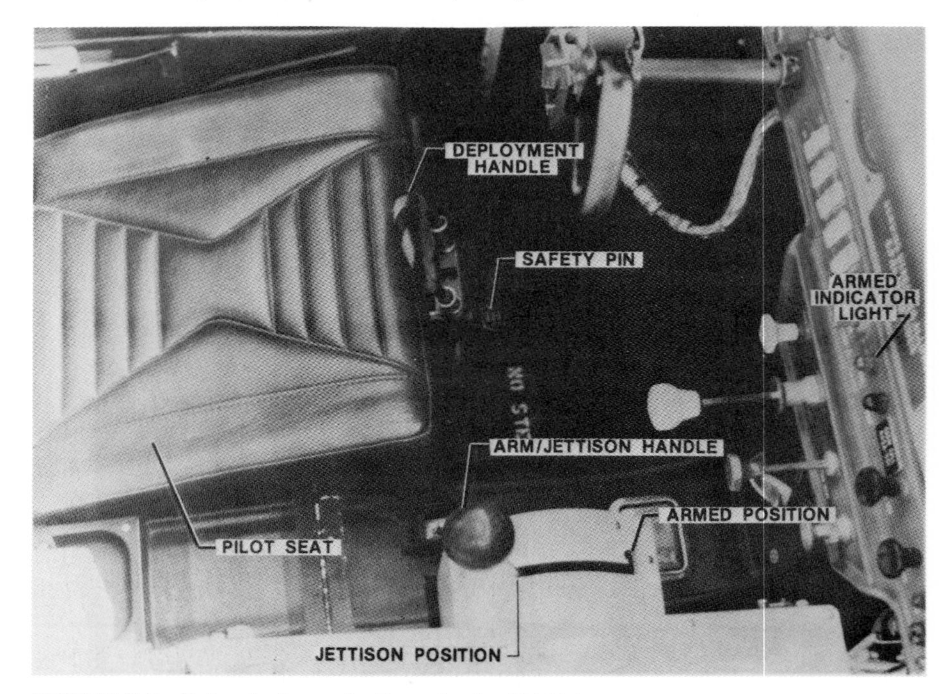

FIGURE 8.3 Spin-chute cockpit controls. *(NASA.)*

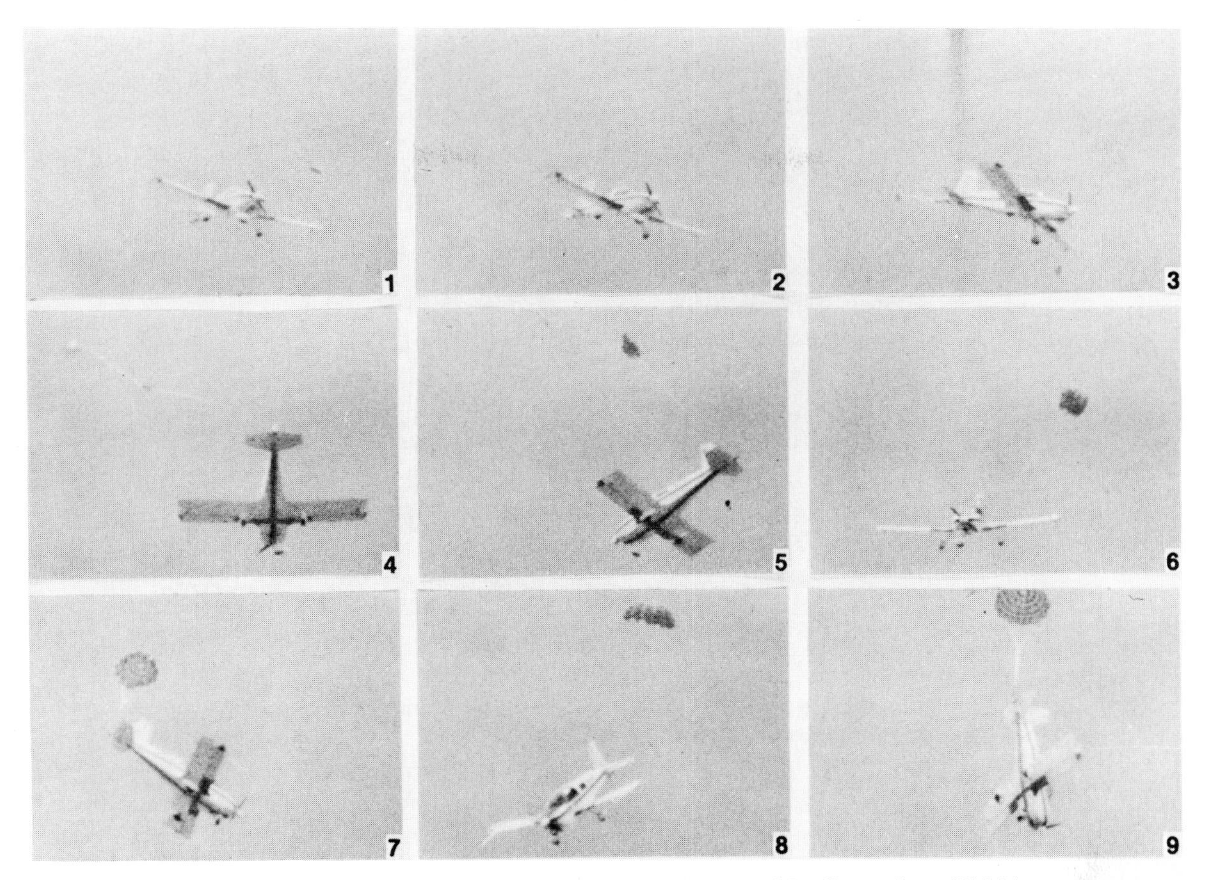

FIGURE 8.4 Utilizing spin chute to stop otherwise unrecoverable flat spin. *(NASA.)*

that nylon can be easily severed when it rubs against something sharp. This once happened while I was using the spin chute to stop a spin in a T-33. The nylon cable severed as it rubbed against the tail pipe. Fortunately, the parachute had already produced enough drag to improve the spin to the point from which I was able to complete the recovery with normal controls.

Two spin-chute controls should be mounted in the cockpit. They should be conveniently located, preferably near the throttle, so that the pilot can grasp them easily under adverse conditions. These controls should look and operate distinctly differently from one another. The most conveniently located one should be the parachute deployment handle. The other should be the control that unlocks the hook that attaches the cable to the airplane. This hook should always be unlocked while taking off or landing. The pilot should always confirm by radio to the chase pilot or someone monitoring a

FIGURE 8.5 Charles Bradshaw, designer of spin recovery chute installation for modified Grumman Yankee.

ground-based radio that the spin chute is unlocked before taking off and before approaching the airport for a landing. The pilot should certainly confirm that it is locked before spinning.

NASA's research confirms that a 10- or 11-ft-diameter parachute fastened to 20 ft of dragline is optimum for stopping a full-blown flat spin in a lightplane. Can you imagine how an airplane would behave if this chute deployed accidentally in flight? The unlocking feature is extremely important.

The cable length is also important. If the cable is too long, the chute will trail well above the spin and will be ineffective in retarding yaw. If it is too short, the wake from the airplane will cause erratic parachute behavior. The size of the canopy and the length of the dragline are determined by the worst possible condition: the flat spin. The flatter the spin, the more parachute area it takes to stop the rotation. Usually, the flatter the spin, the faster the rotational speed. Also, the flatter the spin, the greater the autorotational power. This isn't always true because sometimes flat spins slow down beyond a certain angle of attack, but it has been my experience that it is generally true.

The deployment and jettisoning actuations should both be thoroughly tested on the ground. The unlocking device should work freely while applying a good load against the drag cable. The pilot should not have to pull over 10 lb against the jettisoning handle to unlock the hook with a load against it.

Although spin chutes are used primarily to stop irrecoverable spins, they are also invaluable when a spin may eventually be stopped but the altitude remaining is not sufficient. Don't wait too long before deploying it! Remember, it may take several turns for the parachute to get into an opening position and drag the cable against the restraining hook.

Those who are investigating or consistently performing flat spins would do well to install this emergency recovery device on their airplanes. As a matter of fact, I have long believed that a spin-chute installation would be excellent insurance for aerobatic aircraft, especially when they are being flown by pilots who are first learning aerobatics. The spin chute would not only stop spins, but recover the airplane from *any* altitude.

I watched a movie of an S-1 Pitts spinning inverted into the ground from the top of a hammerhead. If this airplane had had a spin chute, the pilot would probably have survived.

I realize that an expert aerobatic pilot would feel the same as a powerful swimmer who was wearing water wings, but there are times when even experienced pilots need all the help they can get. It wouldn't take a very large chute to snatch some of these small aerobatic airplanes to attention, and it would not change the shape of the airplane to any great extent or affect their aerodynamic characteristics. If an aerobatic airplane was designed to include a spin-chute installation, the container could probably be built within the fuselage contour.

In some jet fighter aircraft, it is a recommended procedure to use the drag chute to assist recoveries from poststall gyrations and spins. The F-4 is one such airplane, although the procedure hasn't proved very effective because a drag chute is not large enough to do the trick.

BAILING OUT

Even though you may jump to the outside of a flat spin, bailing out is about as safe as jumping through a spinning propeller. You're almost certain to be struck by the airplane. Even during inverted flat spins, when you are most likely to clear the airplane, after you have opened your parachute, the airplane is likely to descend into the canopy. For this reason, you should delay opening your parachute for as long as possible.

CONFIGURATION CARE

If the airplane has a proven design that has displayed good spin recovery characteristics, care should be taken to make certain that it is rigged according to specifications, and that all of the control surfaces have the proper amount of deflection. Excessive elevator travel, for example, can result in flatter-than-normal spins.

All of the controls should operate freely, with as little friction as possible. It is impossible to conduct meaningful stability checks, an important part of pre-spin testing, with excessive friction in the control system.

When the c.g. is computed, it should include the weight of the pilot who will be conducting the tests. The c.g. should not only be within limits, it should favor the forward limit for the initial spin work.

If the aircraft configuration or weight has been altered from the original design, the spin characteristics can also change. The airplane should be treated as a prototype as far as the spin investigation is concerned. Sometimes *very* small aerodynamic changes can drastically alter an airplane's spin characteristics.

Many years ago I owned a rather odd-looking airplane called a Barling NB-3. The "NB" could have stood for "No Beauty," for it was one of the ugliest airplanes I have ever seen. It had a very thick, full-cantilever, low wing that started its dihedral very abruptly about halfway out on each wing. The fuselage had a normal appearance until just aft of the rear cockpit. There, it appeared that either the designer ran out of pencil lead or the company ran short of money, because the fuselage suddenly narrowed to hardly anything and at the end was mounted a very small set of tail feathers.

The machine performed and handled reasonably well until I installed a larger windshield. When I had finished, I decided to take the airplane for a short test flight.

About 1500 ft above the airport I kicked the Barling into a spin, intending to spin no lower than pattern altitude. However, I detected a change of spin character almost immediately. I shoved hard rudder against the spin, followed by full forward stick. The spin continued, but as I held the controls in the recovery position, the nose started to drop. I saw objects on the ground become larger through my clear new windshield. Finally the rotation slowed and the spin stopped. The Barling was now flying below a set of power lines close off my right side. I was well below 100 ft!

Soon after landing, I removed my new windshield and put the old one back on. It was apparent that the larger windshield was influencing the airflow over the tail surfaces.

RIGGING

Before starting a spin-test program, the airplane should be flown enough to cure any rigging problems. There should be no wing heaviness or any tendency to yaw when the airplane is flown straight and level at cruising speed.

STABILITY CHECKS If the airplane is a new design, or has been reconfigured from an original design, basic stability checks should be performed. Usually, these tests are conducted throughout the entire speed range of the airplane. However, when considering stalls and spins, it is important to conduct these tests at as slow an airspeed as possible. If the airplane can be trimmed longitudinally (pitch) at approach speed or 1.3 times the 1-*g* stall speed, use that speed; if not, use the minimum trim speed. Use whatever power is necessary to maintain altitude.

With the airplane in trim, displace the nose upward about 10° by applying back pressure against the stick. Then release the stick and count the number of oscillations below and above the horizon to return to level flight and a stable trim speed. The number of oscillations will vary with the location of the c.g. There will be fewer oscillations at forward c.g. than at aft c.g. The important thing is that each successive oscillation becomes less and that an eventual return to the original trim speed takes place. If the oscillations diverge relative to the horizon (become greater and greater), the longitudinal stability is unsatisfactory and the airplane should be redesigned. This may involve increasing the size of the horizontal stabilizer, or relocating it up or down until it is flying in an effective airflow. Sometimes, changing the angle of attack of the stabilizer will help. The stabilizers of stable aircraft carry a down load, which involves setting the surface at a negative angle of attack.

The c.g. on well-behaved airplanes with rear stabilizers is always ahead of the center of pressure (the center of lift). This means that when the wings start to lift, there is a nose-down pitching moment. This moment is corrected by applying negative lift to the horizontal stabilizer. As the airplane increases speed, the negative lift increases, resulting in a nose-up pitching moment. In other words, as airspeed increases the nose tends to rise, requiring a nose-down trim. Conversely, when airspeed is reduced, there is less negative lift on the horizontal stabilizer, resulting in the nose pitching downward, and requiring nose-up trim to hold the airplane level.

It is within this relationship that the c.g. becomes important. As the c.g. is moved aft toward the center of pressure, the stabilizing effect is reduced. When the c.g. is colocated with the center of pressure, the stability becomes neutral. In other words, when you lower the nose, it stays down, when you pull the nose up, it stays up. There is no restoring force to return the airplane to level flight.

When the c.g. is located aft of the center of pressure, the airplane becomes dangerously unstable, particularly at low speeds. During low-speed flight, there will not be enough elevator power to prevent the nose from pitching up, and the airplane may enter an unrecoverable stall or spin.

It has been my observation that very few pilots pay enough attention to where the c.g. is located, and do not appreciate the importance of flying an airplane where the center-of-pressure and center-of-gravity positions are safely related. Flying an airplane out of aft c.g. limits is an open invitation to an inadvertent and unrecoverable spin.

A most important low-speed stability investigation prior to performing stalls and spins is the exploration of the directional, or yaw, stability of the airplane.

Full-rudder sideslips at speeds of 5 or 10 knots prior to the 1-*g* stall speed should be done. This is usually performed by smoothly entering a slip by starting a yaw and maintaining enough opposite aileron to maintain a heading. The rudder and aileron are applied until full rudder travel is obtained. Power is used to maintain altitude. The use of power will also permit maximum airflow past the rudder, resulting in maximum yaw angles. The rudder force should increase with yaw angle. There should be no reduction of rudder force, and certainly no reversal of forces where the rudder stays deflected without pressure against it, requiring considerable pressure against the opposite rudder to restore the yaw to neutral.

The usual fix for airplanes with directional stability problems is to add more area to the vertical fin. When the airplane has been proved to be stable in yaw and pitch at low airspeed, it is time to start a stall investigation.

STALL EXPLORATION

From a safe altitude, approach the first stall at a slow deceleration rate of no more than 1 mile per hour per second. This will assure that the stall will occur at 1 *g*, and give you an accurate check on the indicated stall speed, The pilot should be alert for aerodynamic stall warning. The most ideal situation would be to detect stall buffet during the very early stages of the stall. This would reveal that the root sections of the wing are stalling first, a very desirable characteristic. If the first indication of a stall is a rapid roll-off one way or another, a rather undesirable condition exists. It reveals that the outer portion of the wing is stalling first. This is not only undesirable from the standpoint of the stall, but may predict undesirable spin characteristics. The more stall activity that takes place on the outer portion of the wing, the greater will be the spin rotational speed. The airplane may have other characteristics that will result in being able to stop the spin without difficulty, but the situation should be corrected before the actual spin program is started.

Ideally, the outer portions of a wing should be the last to stall. This can be accomplished by either washing out the wing tips to reduce their angle of incidence, or installing a wedge-shaped stall strip along the leading edge of the inboard portion of the wings. Washing out the outer portion of the wings results in the inboard portions reaching a stalling angle of attack before the outer portions. Installing a stall strip along the leading edge of the root section results in that portion of the wing stalling before the wing tips.

In its spin research program, NASA discovered that when the outer portion of the wing of its spin research airplane was modified by drooping the leading edge and modifying the airfoil, the airplane became highly spin resistant. However, when the same airfoil was extended along the entire wing, it easily entered an uncontrollable flat spin. It is obvious that it is important to stall the root section while keeping the outer portion of the wing unstalled as long as possible.

Don't work the stick back any further than necessary on the first few stalls. Use just enough aft stick to produce a stall. If the airplane does not exhibit any stall abnormalities, the stick can be pulled further aft during successive stalls until, finally, the elevators are fully deflected.

The pilot should be alert for the following undesirable symptons:

1. A lightening or reversal of elevator stick forces.
2. A tendency for the nose to pitch upward rather than downward when the airplane first stalls.
3. A lightening of rudder forces, or uncontrollable yaw.

If the airplane displays any of these characteristics during the stall, the airplane should not be spun until they are corrected.

In order to determine what may be causing these abnormalities, the fuselage aft of the cockpit area and the tail surfaces should be tufted with yarn of contrasting colors. The flow patterns can be observed by the pilot of the chase plane, or by a movie camera mounted near the wing tip and focused on the problem area.

Very often these kinds of problems involve a redesign of the tail, possibly moving the horizontal stabilizer up or down into less disturbed air. Sometimes all that is needed is to restrict the amount of up-elevator travel. The maximum up-elevator travel is usually determined by that needed to flare for a landing at the forward c.g. limit. Excessive up-elevator travel results in an excessive angle of attack and a more violent reaction to the stall.

Sometimes a simple wedge-shaped stall strip mounted on the leading edge of the wing near the root section will do the trick. This will result in an early stall in this area and tend to result in an earlier pitch-down of the nose.

There have been situations where the forward part of the fuselage was lifting excessively, resulting in a pitch-up effect at the stall. This can often be corrected by installing wedge spoilers (strakes) across or around the nose.

These effects can be determined by tufting the forward fuselage and observing their behavior after the wing has stalled. If a considerable forward area of the fuselage reveals that it is lifting, spoilers or a redesign may be necessary. Airplanes with considerable fuselage area ahead of the wing sometimes have this problem.

It is common practice during stall/spin investigations to install a stick restrictor that folds back out of the way for the landing. This prevents the pilot from applying more aft stick than was intended. This stop can be adjustable so that the amount of stick travel can be varied.

AFT C.G. TESTING It should be remembered that as the c.g. is moved aft, it is easier to obtain a high angle of attack, and the amount of up-elevator travel may be an important consideration. All initial stall/spin work should be accomplished at the maximum forward c.g. and moved aft in small increments.

Aft c.g. stall/spin investigations should be accomplished by installing jettisonable lead shot in a container on the aft part of the airplane. The design of the container should contain a quick-jettisoning feature. A small parachute-type pack container with quick-release pins actuated from the cockpit is an ideal way to handle the shot. Sometimes the shot release is fastened to the spin-chute deployment handle so that the shot is released simultaneously with spin-chute deployment. Another method is to jettison the shot during the initial travel of the deployment handle, and to actuate the parachute during the final travel of the handle.

During the initial phase of lightplane spin testing, you should have a minimum ground clearance of 8000 ft. The pilot should predetermine the minimum altitude at which to deploy the spin chute if recovery controls are ineffective in stopping the spin. Without a method of immediately projecting the parachute away from the spin wake, it will take an indeterminate number of rotations before the chute projects far enough into the free airstream to open and drag the rotation to a stop. All of this consumes an incalculable amount of altitude.

DECISION HEIGHT There should also be a decision about the altitude at which to abandon the airplane. Without an ejection seat, this can take more time than is realized. Removing the headset, jettisoning the door, releasing the restraint harness, getting to the door, getting free of the airplane, pulling the rip cord, and waiting for the parachute to open takes time and plenty of altitude. Each airplane has its own set of bailout problems that should be thoroughly thought through before the need arises.

The pilot should always leave the airplane on the outside of the spin. Doors are usually located only on one side. Provision should be made for going through a window in the event the door is on the inside of the spin.

Although successful bailouts have been made at considerably lower altitudes, 4000 ft above the ground is a safe target altitude for initiating the bailout procedures. The actual exit from the airplane will occur considerably below that altitude.

INITIAL SPIN EXPOSURE

The first spins should be conducted with the landing gear and flaps retracted and at the maximum forward c.g.

Spins are more effectively entered a few miles per hour above the 1-*g* stalling speed. This is because the elevators and rudder are flying in relatively undisturbed air at that speed and are more effective in producing an entry. Often, a slight amount of power in propeller-driven aircraft will provide the additional airflow over the elevator and rudder to effect a better spin entry.

If power is used, it should be reduced to idle the moment autorotation starts. It is a good idea to conduct an idle rpm check at a speed just above the stall prior to doing spins. Spins have a way of disturbing the carburetion, and if the engine is idling too slowly, it is more likely to quit during the spin.

The first exposure to spins should involve rotations of between one-half and one turn. Use only enough aft stick to excite the spin; do not continue to pull the stick aft once autorotation has started. Be particularly aware of excessive rotational speed or any tendency for the spin to flatten. Often, the first turn of rotation may seem a little flatter than normal, but the second turn usually results in a more nose-down attitude. If the elevator forces are heavy, there is probably no reason to be concerned during the first turn. But if the nose stays less than 45° below the horizon after the first turn, an immediate recovery should be made.

If the nose attitude is higher than what you are used to in airplanes approved for spins, if the rotational speed is excessively fast, or if there is lightening of elevator stick forces, the spin should be stopped with hard opposite rudder and full forward stick. Hold the controls in this position until the spin stops.

It is always better to stop the spin immediately upon the appearance of any one of these three indications. Fly back to the airport, talk to your chase pilot, and look at the films. Think about it for a while before you continue with the investigation. The thinking done on the ground is far superior to airborne cogitation. You may want to make a change in configuration before you proceed.

Frequently, things happen so fast during a spin that proper analysis during autorotation is impossible. Take your time! A few hours of delay may add years to your flying career.

TAKE SMALL STEPS

Every change in configuration demands the same initial caution. Be especially careful as you work toward the aft limits! Eventually you might want to try some spins with the landing gear extended or the flaps down.

When considering spins with the landing gear extended, consider the possible change of c.g. when the gear comes down. If the c.g. moves aft with the gear extended, you should probably do the first tests with the airplane ballasted to compensate for it. If you should experience difficulty in stopping the spin, retracting the landing gear will move the c.g. well forward and aid the recovery. Conversely, if the c.g. moves forward when the landing gear is extended, it might be worth extending the gear when encountering a spin that is difficult to recover from.

DENSITY ALTITUDE

Density altitude is an important consideration during the initial phase of spin testing. It is far better to select a sea-level base so that at test altitude you will experience minimum rotational speed. For example, 8000 ft of ground clearance at Denver will place you at an indicated 13,280 ft. On a warm day, you could easily be at a density altitude of over 15,000 ft. You can expect considerably faster rotational speeds and flatter spins under these conditions.

The safest way to start spin testing is at the lowest possible density altitude with the maximum ground clearance. A cool day at 8000 ft ground clearance above sea level is a good altitude at which to start a spin program in a light airplane.

There is no program that requires more planning and careful analysis than testing an airplane for spin characteristics. However, even under the most conscientious planning, a spin-test pilot should always be ready for the unexpected. I cannot overemphasize the value of a good spin-chute installation. I have used them, and I wouldn't be alive today if it had not been for their ability to halt an unrecoverable spin.

SUMMARY

1. Spin testing of prototype models involves a methodical, step-by-step program.
2. Spin testing is one of the most hazardous areas of flight testing.

3. Good test pilots work at their own pace and do not permit people or circumstances to hurry them.

4. A skilled chase pilot, motion-picture coverage, and good radio communication are all essential.

5. An auxiliary spin-recovery device is absolutely essential.

6. If an airplane is of a proven design that has displayed good spin-recovery characteristics, a check to see if the airplane meets previous standards may be all that is necessary. However, the rigging and the amount of control deflection should be checked to see if they meet specifications. Small departures from the original aerodynamic configurations can often change the spin characteristics. A change in weight can also affect the spin. A weight-and-balance check is essential.

7. Stability checks are an important part of pre-spin testing. It is also important that the airplane display no reversal of control forces or instability during the stall.

8. Start the actual spin investigation with a half-turn spin, then a one-turn spin, and gradually increase the number of rotations. Take your time. Do not attempt too much in any one flight.

THE RESEARCH FRONTIER
The Story of NASA's Spin Research

To those interested in the dynamics of flight and in the research into the design of more controllable and efficient aircraft, the NASA Langley Research Center in Hampton, Virginia, is the most fascinating place on earth. There is a great concentration of knowledge, talent, and skills working together toward various research goals in this superb facility.

During the 1930s and 1940s, the forerunner of NASA, the National Advisory Committee for Aeronautics, conducted in-depth studies of the typical airplanes of that era. Data obtained by this research were significant and informative to designers of both military and general aviation aircraft of that day.

After World War II, spin research on general aviation aircraft took a back seat to research on military aircraft, with its drastically changing configurations. The new designs produced a new set of stall/spin problems that demanded the attention of researchers. There weren't enough trained workers, and there wasn't enough money available to continue the much-needed research toward improving the stall/spin characteristics of general aviation aircraft.

However, in the early 1970s phasing out of military aircraft programs resulted in a modest reduction in required support. A few interested researchers at NASA began to turn their attention to the general aviation area.

In 1973 a relatively small project was initiated to determine the validity of previous data, especially the controversial tail-design criterion. This project was based on model and flight tests of one general aviation aircraft employing a number of different tail configurations. Nine different tail con-

Jim Bowman, Jim Patton, and Sanger Burk. *NASA.)*

Modified Grumman Yankee in 30 × 60 ft wind tunnel. *(NASA.)*

figurations were tested in the spin tunnel, and five were tested by radio-controlled models. As the interest of the research community and the general aviation community grew, several NASA advisory groups and the General Aviation Manufacturers Association (GAMA) pressed for an increase in the level of effort and the scope of the program.

The current NASA general aviation stall/spin program is a major effort directed at producing the technology required for designing airplanes that are inherently safer in regard to stalls and spins. It is of interest to note that the expansion and scope of the program has been modeled after the highly successful military program. The program encompasses a large matrix of testing techniques, facilities, and analytical studies.

The Langley spin tunnel is used to study the developed spin and spin recovery. The spin-recovery parachute geometry requirements for spin tests are also determined. A 1/11-scale model is launched, Frisbee-like, into a vertical stream of air, the upward velocity of which is adjusted to match the model's rate of descent. The model is thus maintained in a perpetual spin until recovery is initiated. I had the opportunity to watch this extremely interesting operation.

NASA also has a rotary balance device that is incorporated with the spin tunnel. Models are spun at predetermined rates and angles to measure forces and moments on various configurations and component effects under spin conditions. Data are then computer-processed to identify equilibrium spin modes and effects of control deflections.

One of the more ingenious areas of NASA's spin research program is the radio-controlled model airplane program. The models duplicate the configurations to be tested on the full-scale aircraft. They are from 1/6 to 1/5 scale, and weigh between 12 and 16 lb. Ingenious miniaturized on-board instrumentation, including a telemetering transmitter to transmit data for recording on an audio cassette recorder, was developed by NASA. Postflight processing converts the recorded data into time histories of the parameters describing the spins. The rapid motions of the models can then be studied in detail. The radio-controlled models can be used to study stall, departure, steady spin, and the spin recovery.

Currently, four typical light, single-engine airplanes are being spun to provide real world data for correlation with model test results.

In just one of NASA's research airplanes, a modified Grumman Yankee, a total of 560 spins have been performed for a total of 3700 turns. Many of these spins were of the type that were flat and unrecoverable. Spin chutes have been used a number of times to stop spins in this airplane.

None of the research aircraft are spun in the production configuration, but are experimentally modified for the testing of a variety of configurations. In addition to a spin-recovery parachute, a Beechcraft Sundowner is

Model of Grumman Yankee being spun in NASA's spin tunnel. *(NASA.)*

Dave Robelen and radio-controlled spin model. *(NASA.)*

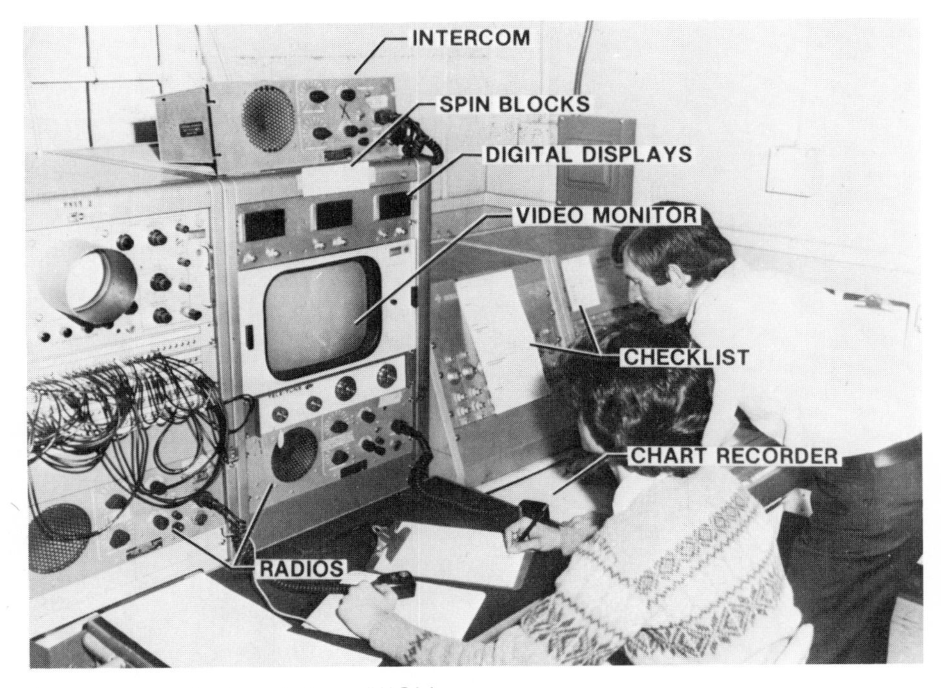

Spin test monitoring station. *(NASA.)*

equipped with throttleable hydrogen peroxide rockets that can be used for either increasing spin rotation rates or stopping spins.

Other research aircraft used in the program are a Cessna 172 Skyhawk and a prototype Piper T tail. All of the aircraft are equiped with wing-tip booms which have swivel heads that provide accurate airspeed, sideslip, and angle-of-attack information. Each airplane is also equipped with external cameras, internal instrumentation, and ballasting provisions to vary the mass distribution and the c.g.

Numerous spin tests have been conducted using a variety of airfoil, tail, and fuselage shapes. The results show that the spin and recovery characteristics of a particular airplane are not necessarily dependent on any single factor, but, rather, may be dependent on the way the parts are put together. Often, spin characteristics can be drastically changed by what may seem like a minor configuration change. One factor can overpower other variables.

NASA has a 12 ft and a 30 $\times$ 60 ft wind tunnel to further explore the aerodynamics of these configurations and to make detailed explorations of small configuration changes.

My association with NASA's Langley Research facility has left me with the assurance that our tax dollars are being efficiently used by a group of enthusiastic, dedicated individuals.

Rotary balance device. *(NASA.)*

Cockpit instrumentation: Modified Grumman Yankee. *(NASA.)*

One-third scale free flight test model. *(NASA.)*

Veteran spin research pilot Jim Patton. *(NASA.)*

THE CONCLUSION IS YOURS

An airplane is an expression of freedom. It permits humans to ascend heavenward before they permanently leave their earthly station behind. However, a lack of training and knowledge can expedite their permanent departure.

Flying might be compared to a bug in a bun. It's perfectly safe until it's exposed to the teeth of the unexpected. The unexpected might be expressed as the sinister shadow that is reflected by complacent familiarity. The feel of controls that always respond, the soothing sound of an engine that has never failed, and the accumulated hours of uneventful flying that magnify each moment of deceptive intimacy provide a very fickle fortress against any deviation from characteristic routine.

If we spend enough time in a rocking chair, we become incapable of reacting properly to even the minor disturbances of life. Vital living demands a constant expansion of the abilities of mind and body. So it is with flying. Real flying skills are developed through exercises that constantly stretch our abilities.

Even the simplest of flying tasks can be performed better if we have been involved in more complex exercises. For example, those who have been exposed to aerobatics feel perfectly comfortable in turbulence that might unnerve the pilot who never exceeds 30° of bank.

You never know when extra flying skills will come in handy. A friend of mine, Johnny Miller, suddenly found himself with a jammed elevator control after a midair collision. As the nose of the airplane pitched upwards, he rolled the airplane into a steep bank and, by the use of top rudder, entered a severe sideslip. He was able to control the nose attitude by the use of top and bottom rudder. When he hit the ground in a dry riverbed, the low wing absorbed most of the shock, and he walked away from the wreckage. I wonder how many pilots know that you can control nose attitude by rolling into a slip and using the rudder in place of the elevators? Johnny exhibited a really slick bit of flying, but he was ready for the unexpected because he had exercised his skills beyond that which is normally expected.

Just practicing stalls and spins will not provide all of the immunization you may need against abnormal flight behavior. A course in basic aerobatics is an excellent insurance policy against unusual and unexpected attitudes.

Flying is fun, but many pilots miss out on the greatest satisfaction of flying, and that is accepting the challenge of increasing and maintaining flying skills. Those who maintain only limited skills will be constantly haunted by the knowledge that they may suddenly and unexpectedly be called upon to do more than they are capable of. This fear is an unnecessary burden for pilots to bear. It can be eliminated with a little effort.

Why not resolve to enter into a flying fitness program today? Your family will love you longer because of your decision.

GENERAL AVIATION STALL/SPIN BIBLIOGRAPHY

Adams, W. M., Jr.: *SPINEQ: A Program for Determining Aircraft Equilibrium Spin Characteristics Including Stability,* NASA TM-78759, 1978.

Agrawal, Shreekant: *Parametric Studies of Discontinuous Leading-Edge Effects on Wing Characteristics for Angles of Attack up to 50°*, Master of Science thesis, University of Maryland, August 1979.

Anderson, Seth B.: *A Historical Overview of Stall/Spin Characteristics of General Aviation Aircraft,* AIAA Conference on Air Transportation: Technical Perspectives and Forecasts, August 1978.

Bement, Laurence J.: *Emergency In-Flight Egress Opening for General Aviation Aircraft,* NASA TM 80235, April 1980.

——: *Emergency In-Flight Egress Opening for General Aviation Aircraft,* NASA CP-2127, 14th Aerospace Mechanisms Symposium, May 1980.

——: *Emergency In-Flight Egress Opening for General Aviation Aircraft,* paper presented at 18th Annual SAFE Symposium, October 1980.

Bennett, A. G., and **J. K. Owens:** *Investigation of a Stall Deterrent System Utilizing an Acoustic Stall Sensor,* SAE Paper 770473, presented at the SAE Business Aircraft Meeting, Wichita, Kan., March 1977.

——, ——, and **G. Ball:** *A Study of Stall Deterrent Systems for General Aviation Aircraft,* AIAA Paper No. 80-1562.

Bihrle, William, Jr., and **James S. Bowman, Jr.:** *The Influence of Wing, Fuselage, and Tail Design on Rotational Flow Aerodynamics Data Obtained Beyond Maximum Lift with General Aviation Configurations,* AIAA Paper No 80-0455, AIAA 11th Aerodynamic Testing Conference, March 1980.

Bihrle Associates: *Rotary Balance Data for a Single-Engine Trainer Design for an Angle-of-Attack Range of 8 to 90°*, NASA CR-3099, August 1979.

This bibliography is based on material prepared by Paul Stough, NASA, Langley Research Center, Hampton, VA 23665.

——: *Rotary Balance Data for a Typical Single-Engine General Aviation Design for an Angle-of-Attack Range of 30 to 90°*, NASA CR-2972, July 1978.

——: *Rotary Balance Data for a Typical Single-Engine General Aviation Design for an Angle-of-Attack Range of 8 to 90°: I—High-Wing Model B,* NASA CR-3097, September 1979.

——: *Rotary Balance Data for a Typical Single-Engine General Aviation Design for an Angle-of-Attack Range of 8 to 90° II—Low-Wing Model B,* NASA CR-3098, September 1979.

——: *Rotary Balance Data for a Typical Single-Engine General Aviation Design for an Angle-of-Attack Range of 8 to 90° II—High-Wing Model A,* NASA CR-3101, September 1979.

——: *Rotary Balance Data for a Typical Single-Engine General Aviation Design for an Angle-of-Attack Range of 8 to 90° II—High-Wing Model C,* NASA CR-3201, November 1979.

——: *Rotary Balance Data for a Typical Single-Engine General Aviation Design for an Angle-of-Attack Range of 8 to 35° III—Effect of Wing Leading-Edge Modifications Model A,* NASA CR-3102, November 1979.

——: *Rotary Balance Data for a Typical Single-Engine General Aviation Design for an Angle-of-Attack Range of 8 to 90° I—Low-Wing Model A,* NASA CR-3100, February 1980.

——: *Rotary Balance Data for a Typical Single-Engine General Aviation Design for an Angle-of-Attack Range of 8 to 90° I—Low-Wing Model C,* NASA CR-3200 (in publication).

——: *Rotary Balance Data for a Typical Single-Engine General Aviation Design Having a High Aspect-Ratio Canard for an Angle-of-Attack Range of 8 to 90°,* NASA CR-3170 (in review).

——: *Static Aerodynamic Characteristics of a Typical Single-Engine Low-Wing General Aviation Design for an Angle-of-Attack Range of −8 to 90°*, NASA CR-2971, July 1978.

Bowman, James S., Jr.: *Summary of Spin Technology as Related to Light General Aviation Airplanes,* NASA TN D-6575, December 1971.

——, **Harry P. Stough, Sanger M. Burk, Jr.,** and **James M. Patton, Jr.:** *Correlation of Model and Airplane Spin Characteristics for a Low-Wing General Aviation Research Airplane,* AIAA Paper No. 78-1477, August 1978.

Bradshaw, Charles F.: *A Spin-Recovery Parachute System for Light General Aviation Airplanes,* NASA CP-2127, 14th Aerospace Mechanisms Symposium, May 1980.

——, and **H. P. Stough:** *Design and Development of a Spin-Recovery Parachute System for Light General Aviation Aircraft,* paper presented at SAFE Association 17th Annual Symposium, December 1979.

Burk, Sanger M., Jr., James S. Bowman, Jr., and **William L. White:** *Spin-Tunnel Investigation of the Spinning Characteristics of Typical Single-Engine General Aviation Designs, I—Low-Wing Model A: Effects of Tail Configurations,* NASA TP-1009, September 1977.

——, ——, and ——: *Spin-Tunnel Investigation of the Spinning Characteristics of Typical Single-Engine General Aviation Designs, I—Low-Wing Model A: Tail Parachute Diameter and Canopy Distance for Emergency Spin Recovery,* NASA TP-1076, November 1977.

——, and **Calvin F. Wilson, Jr.:** *Radio-Controlled Model Design and Testing Techniques for Stall/Spin Evaluation of General Aviation Aircraft,* NASA TM-80510, 1975.

Chambers, Joseph R.: *Overview of Stall/Spin Technology,* AIAA Paper 80-1580, August 1980.

Chevalier, Howard L.: *Aerodynamic Concept for Stall Proofing a General Aviation Airplane,* Grant NSG 1407, Texas A&M University, October 1978.

——: *Some Theoretical Considerations of a Stall Proof Airplane,* SAE Paper No. 790604, SAE Business Aircraft Meeting, April 3–6, 1979.

——, and **Joseph C. Brusse:** *A Stall/Spin Prevention Device for General-Aviation Aircraft,* SAE paper No. 730333, SAE Business Aircraft Meeting, Wichita, Kan., April 3–6, 1973.

——, **Robert A. Wilke,** and **Michael L. Faulkner:** *Wind-Tunnel Evaluation of Aerodynamic Spoiler on General Aviation Aircraft Horizontal Tail, Part I,* Grant NSG 1407, Texas A&M University, May 1979.

DiCarlo, Daniel J., and **Joseph L. Johnson, Jr.:** *Exploratory Study of the Influence of Wing Leading-Edge Modifications on the Spin Characteristics of a Low-Wing Single-Engine General Aviation Airplane,* AIAA Paper No. 79-1837, AIAA Aircraft Systems and Technology Meeting, New York, August 1979.

——, **H. P. Stough, 3d,** and **J. M. Patton, Jr.:** *Effects of Discontinuous Drooped Wing Leading-Edge Modifications on Spinning Characteristics of a Low-Wing General Aviation Airplane,* AIAA Paper No. 80-1843, August 1980.

Feistel, T. W., S. B. Anderson, and **R. A. Kroeger:** *A Method for Localizing Wing Flow Separation at Stall to Alleviate Spin Entry Tendencies,* AIAA Paper 78-1476, AIAA Aircraft Systems and Technology Conference, August 1978.

Johnson, J. L., W. A. Newson, and **D. R. Satran:** *Full-Scale Wind-Tunnel Investigation of the Effects of Wing Leading-Edge Modifications on the High Angle-of-Attack Aerodynamic Characteristics of a Low-Wing General Aviation Airplane,* AIAA paper No. 80-1844, August 1980.

Joyner, Jerry Wayne: *Estimation of Logitudinal Stability Derivatives Using Steady State Flight Test Data,* Master of Science thesis, Mississippi State University, April 1977.

King, Michael L.: *A Digital Controller for an Active Stall Deterrent Control System,* Master of Science thesis, Mississippi State University, May 1978.

Klein, Vladislav: *Determination of Stability and Control Parameters of a Light Airplane From Flight Data Using Two Estimation Methods,* NASA TP-1306, March 1979.

—— and **James R. Schiess:** *Compatibility Check of Measured Aircraft Responses Using Kinematic Equations and Extended Kalman Filter,* NASA TN D-8514, August 1977.

Kroeger, R. A.; and **T. W. Feistel:** *Reduction of Stall-Spin Entry Tendencies Through Wing Aerodynamic Design,* SAE Paper 760481, April 1976.

Moul, Thomas M.: *Wind-Tunnel Investigation of the Flow Correction for a Model-Mounted Angle-of-Attack Sensor at Angles of Attack from −10° to 110°* NASA TM-80189, November 1979.

—— and **Lawrence W. Taylor, Jr.:** *Determination of an Angle-of-Attack Sensor Correction for a General Aviation Airplane at Large Angles-of-Attack as Determined from Wind-Tunnel and Flight Tests,* AIAA Paper No. 80-1845, AIAA Systems and Technology Meeting, Anaheim, Calif., August 1980.

Nguyen, Luat T.: *Control System Techniques for Improved Departure/Spin Resistance for Fighter Aircraft,* paper No. 791083, SAE Aerospace Meeting, December 3–6, 1979.

O'Bryan, Tom, Ken Glover, and **Tom Edwards:** *Some Results from the Use of a Control Augmentation System to Study the Developed Spin of a Light Plane,* AIAA Paper No. 79-1790, AIAA Aircraft Systems and Technology Meeting, August 1979.

O'Bryan, T. C., M. W. Goode, F. D. Gregory, and **M. H. Mayo:** *Description of Experimental (Hydrogen Peroxide) Rocket System and its Use in Measuring Aileron and Rudder Effectiveness of a Light Airplane,* NASA TP-1647, May 1980.

Patton, James, M., Jr.: *General Aviation Stall/Spin Research Program,* Paper presented at Society of Experimental Test Pilots, European Section, 1979 Symposium, Bristol, England, April 1979.

——, **H. Paul Stough, 3d,** and **Daniel J. DiCarlo:** *Spin Flight Research Summary,* SAE Paper No. 790565, SAE Business Aircraft Meeting and Exposition, Wichita, Kan., April 1979.

Saini, Jugal Kishore: *An Experimental Investigation of the Effects of Leading-Edge Modification on the Post-Stall Characteristics of an NACA 0015 Wing,* Master of Science thesis, University of Maryland, August 1979.

Simpson, Don E.: *An Experimental Investigation of the Output Characteristics of an Acoustic Stall Sensor,* Master of Science thesis, Mississippi State University, August 1975.

Sliwa, Steven M.: *Some Flight Data Extraction Techniques Used on a General Aviation Spin Research Aircraft,* AIAA Paper No. 79-1802, AIAA Aircraft Systems and Technology Meeting, Flight Tests Methods Session, August 1979.

——: *A Study of Data Extraction Techniques for Use in General Aviation Aircraft Spin Research,* Master of Science thesis, George Washington University, September 1978.

Staff of Langley Research Center: *Exploratory Study of the Effects of Wing Leading-Edge Modifications on the Stall/Spin Behavior of a Light General Aviation Airplane,* NASA TP-1589, December 1979.

Stough, H. P., 3d, D. J. DiCarlo, and **J. M. Patton, Jr.:** *Spinning for Safety's Sake,* Paper presented at SAFE Association 17th Annual Symposium, December 1979.

——, and **James M. Patton, Jr.:** *The Effects of Configuration Changes on Spin and Recovery Characteristics of a Low-Wing General Aviation Research Airplane,* AIAA Paper No. 79-1786, AIAA Aircraft Systems and Technology Meeting, New York, August 1979.

Taylor, Lawrence W., Jr., and **Kenneth W. Iliff:** *Systems Identification Using a Modified Newton-Raphson Method—A Fortran Program,* NASA TN D-6734, May 1972.

Tischler, M. B., and **J. B. Barlow:** *Application of the Equilibrium Spin Technique to a Typical Low-Wing General Aviation Design,* AIAA Paper No. 19-1625, AIAA Atmospheric Flight Mechanics Conference, Boulder, Colo., August 1979.

—— and ——: *Determination of the Spin and Recovery Characteristics of a Typical Low-Wing General Aviation Design,* AIAA Paper No. 80-0169, 1980.

Tischler, Mark B.: *Equilibrium Spin Analysis with an Application to a General Aviation Design,* Master of Science thesis, Dept. of Aerospace Engineering, University of Maryland, College Park, Md., October 1979.

Toulmay, Francois: *Analytical Investigation of Several Actuator Concepts for Stall Deterrent Systems,* Master of Science thesis, Mississippi State University, August 1977.

Winkelmann, A. E., and **J. B. Barlow:** *The Effect of Aspect Ratio on Oil Flow Patterns Observed on a Wing Beyond Stall,* TR AE -79-4, Dept. of Aerospace Engineering, University of Maryland, College Park, Md., December 1979.

——, ——, **J. K. Saini, J. D. Anderson, Jr.,** and **E. Jones:** *The Effects of Leading-Edge Modifications on the Post-stall Characteristics of Wings,* AIAA 18th Aerospace Sciences Meeting, January 1980.

INDEX

ABOUT THE AUTHOR

Sammy (Samuel H.) Mason soloed an OX-5 - powered Eaglerock biplane at the age of 16. He was a flight instructor during World War II, organized the Hollywood Hawks air show group, and was considered the nation's foremost aerobatic pilot during the late 1940s.

He retired from Lockheed Aircraft Corporation in 1973 after 23 years as an engineering test pilot. He was the first pilot to do extensive aerobatics in a helicopter and performed in Lockheed's 286 Rigid Rotor helicopter at the 1968 Paris Air Show.

Always interested in flight training, he was given the 1976 Flight Instructor of the Year award by the National Association of Flight Instructors for new instructional techniques. He is still an active flight instructor and specializes in aerobatic and safety training.